Functional and Virtual Medical 3D Imaging in Cardiovascular Medicine

Education of the Novel Medical Approaches in Daily Clinical Practice

Functional and Virtual Medical 3D Imaging in Cardiovascular Medicine

Education of the Novel Medical Approaches in Daily Clinical Practice

Edited by

Artur Rydosz

Biomarkers Analysis LAB, Institute of Electronics, AI, AGH University of Kraków, Kraków, Poland

Klaudia Proniewska-van Dam

Center for Digital Medicine and Robotics, Jagiellonian University Medical College, Kraków, Poland; Functional and Virtual Medical 3D Imaging Laboratory, Department of Diagnostic Imaging, University Hospital in Kraków, Kraków, Poland

Peter van Dam

Department of Radiology, Jagiellonian University Medical College, Kraków, Poland; Functional and Virtual Medical 3D Imaging Laboratory, Department of Diagnostic Imaging, University Hospital in Kraków, Kraków, Poland

ELSEVIER

ACADEMIC PRESS

An imprint of Elsevier

Academic Press is an imprint of Elsevier
125 London Wall, London EC2Y 5AS, United Kingdom
50 Hampshire Street, 5th Floor, Cambridge, MA 02139, United States

ISBN: 978-0-443-44570-5

For Information on all Academic Press publications
visit our website at https://www.elsevier.com/books-and-journals

Publisher: Megan Ball
Acquisitions Editor: Elizabeth Brown
Editorial Project Manager: Shubham Dixit
Production Project Manager: Vasantha Gokilam
Cover Designer: Raman Kumar

Typeset by MPS Limited, Chennai, India

Contents

List of contributors

Paweł Banyś
Functional and Virtual Medical 3D Imaging Laboratory, Department of Diagnostic Imaging, University Hospital in Kraków, Kraków, Poland

Emanuele Barbato
Department of Clinical and Molecular Medicine, Sapienza University of Rome, Rome, Italy; Cardiology Unit, Sant'Andrea University Hospital, Rome, Italy

Rogier Barendse
Department of Cardiology, Erasmus MC (Erasmus University Medical Center), Rotterdam, Netherlands

Marta Belmonte
Department of Clinical and Molecular Medicine, Sapienza University of Rome, Rome, Italy; Cardiology Unit, Sant'Andrea University Hospital, Rome, Italy

Bo Berends
Radboudumc 3D Lab, Radboud University Medical Center, Nijmegen, The Netherlands

Łukasz Brandt
Functional and Virtual Medical 3D Imaging Laboratory, Department of Diagnostic Imaging, University Hospital in Kraków, Kraków, Poland; Department of Electroradiology, Jagiellonian University Medical College, Kraków, Poland

Nico Bruining
Department of Cardiology, Erasmus MC (Erasmus University Medical Center), Rotterdam, Netherlands

Francesco Burzotta
Fondazione Policlinico Universitario A. Gemelli IRCCS, Università Cattolica del Sacro Cuore, Rome, Italy

Gianluca Campo
Interventional Cath-Lab, University of Ferrara, Ferrara, Italy

Bernadeta Chyrchel
2nd Department of Cardiology, Jagiellonian University Medical College, Kraków, Poland; Department of Diagnostic Imaging, University Hospital in Kraków, Kraków, Poland

Antoni Cierniak
Center for Digital Medicine and Robotics, Jagiellonian University Medical College, Kraków, Poland

Dylan Duits
Radboudumc 3D Lab, Radboud University Medical Center, Nijmegen, The Netherlands

Julianna Dąbrowa
Center for Digital Medicine and Robotics, Jagiellonian University Medical College, Kraków, Poland

Roberto Ferrari
Interventional Cath-Lab, University of Ferrara, Ferrara, Italy

Kinga Glądys
Center for Digital Medicine and Robotics, Jagiellonian University Medical College, Kraków, Poland

Grzegorz Goncerz
Department of Anatomy, Jagiellonian University Medical College, Kraków, Poland

Michał Goncerz
Department of Anatomy, Jagiellonian University Medical College, Kraków, Poland

Dominik Grochala
Biomarkers Analysis LAB, Institute of Electronics, AGH University of Kraków, Kraków, Poland

Anna Grochowska
Department of Radiology, Jagiellonian University Medical College, Kraków, Poland

Paweł Kacprzak
Advanced Medical Equipment, Warsaw, Poland; Helgelandssykehuset HF, Mo I Rana, Norway

Stanisław Karcz
Biomarkers Analysis LAB, Institute of Electronics, AGH University of Kraków, Kraków, Poland

Michał Karny
Center for Digital Medicine and Robotics, Jagiellonian University Medical College, Kraków, Poland

Mateusz Kocoń
Biomarkers Analysis LAB, Institute of Electronics, AGH University of Kraków, Kraków, Poland

Julia Kolasa
Center for Digital Medicine and Robotics, Jagiellonian University Medical College, Kraków, Poland; Functional and Virtual Medical 3D Imaging Laboratory, Department of Diagnostic Imaging, University Hospital in Kraków, Kraków, Poland

Jan Kost
Center for Digital Medicine and Robotics, Jagiellonian University Medical College, Kraków, Poland; Functional and Virtual Medical 3D Imaging Laboratory, Department of Diagnostic Imaging, University Hospital in Kraków, Kraków, Poland

Maciej Lis
Department of Anatomy, Jagiellonian University Medical College, Kraków, Poland

Peter Loh
Department of Cardiology, UMC Utrecht, Utrecht, the Netherlands

Thomas Maal
Radboudumc 3D Lab, Radboud University Medical Center, Nijmegen, The Netherlands

Szymon Maciaszek
Center for Digital Medicine and Robotics, Jagiellonian University Medical College, Kraków, Poland; Functional and Virtual Medical 3D Imaging Laboratory, Department of Diagnostic Imaging, University Hospital in Kraków, Kraków, Poland

Krzysztof Piotr Malinowski
Center for Digital Medicine and Robotics, Jagiellonian University Medical College, Kraków, Poland; Functional and Virtual Medical 3D Imaging Laboratory, Department of Diagnostic Imaging, University Hospital in Kraków, Kraków, Poland

Paweł T. Matusik
Department of Electrocardiology, Institute of Cardiology, Jagiellonian University Medical College, Kraków, Poland; Department of Electrocardiology, St. John Paul II Hospital, Kraków, Poland

Michał Mączewski
Department of Clinical Physiology, Centre of Postgraduate Medical Education, Warsaw, Poland

Grzegorz J. Nalepa
Jagiellonian Human-Centered Artificial Intelligence Department (JAHCAI), Institute of Applied Computer Science, Jagiellonian University, Kraków, Poland

Anna Paleczek
Biomarkers Analysis LAB, Institute of Electronics, AGH University of Kraków, Kraków, Poland

Michał Piotrowski
Center for Digital Medicine and Robotics, Jagiellonian University Medical College, Kraków, Poland

Tadeusz J. Popiela
Department of Radiology, Jagiellonian University Medical College, Kraków, Poland

Klaudia Proniewska-van Dam
Center for Digital Medicine and Robotics, Jagiellonian University Medical College, Kraków, Poland; Functional and Virtual Medical 3D Imaging Laboratory, Department of Diagnostic Imaging, University Hospital in Kraków, Kraków, Poland

Miłosz Rozynek
Department of Diagnostic Imaging, University Hospital in Kraków, Kraków, Poland; Department of Radiology, Jagiellonian University Medical College, Kraków, Poland

Robert Runiński
Department of Diagnostic Imaging, University Hospital in Kraków, Kraków, Poland; Diagnostic and Interventional Radiology Unit, BIOMORF Department, University Hospital Messina, Messina, Italy

Artur Rydosz
Advanced Medical Equipment, Warsaw, Poland; Biomarkers Analysis LAB, Institute of Electronics, AGH University of Kraków, Kraków, Poland; Functional and Virtual Medical 3D Imaging Laboratory, Department of Diagnostic Imaging, University Hospital in Kraków, Kraków, Poland

Ksenia Sedova
Department of Cardiology, Institute for Clinical and Experimental Medicine, Prague, Czech Republic

Bernard Solewski
Department of Anatomy, Jagiellonian University Medical College, Kraków, Poland

Michał Szafarczyk
Center for Digital Medicine and Robotics, Jagiellonian University Medical College, Kraków, Poland; Functional and Virtual Medical 3D Imaging Laboratory, Department of Diagnostic Imaging, University Hospital in Kraków, Kraków, Poland

Kaja Tatoń
Center for Digital Medicine and Robotics, Jagiellonian University Medical College, Kraków, Poland; Functional and Virtual Medical 3D Imaging Laboratory, Department of Diagnostic Imaging, University Hospital in Kraków, Kraków, Poland

William Uribe
Department of Cardiology and Electrophysiology, CES University, Medellin, Colombia; Center for Digital Medicine and Robotics, Jagiellonian University Medical College, Kraków, Poland

Peter van Dam
Department of Radiology, Jagiellonian University Medical College, Kraków, Poland; Functional and Virtual Medical 3D Imaging Laboratory, Department of Diagnostic Imaging, University Hospital in Kraków, Kraków, Poland

Piotr Walecki
Department of Bioinformatics and Telemedicine, Jagiellonian University Medical College, Kraków, Poland

Sandra Zarychta
Center for Digital Medicine and Robotics, Jagiellonian University Medical College, Kraków, Poland; Functional and Virtual Medical 3D Imaging Laboratory, Department of Diagnostic Imaging, University Hospital in Kraków, Kraków, Poland

Robert Żuk
Center for Digital Medicine and Robotics, Jagiellonian University Medical College, Kraków, Poland; Functional and Virtual Medical 3D Imaging Laboratory, Department of Diagnostic Imaging, University Hospital in Kraków, Kraków, Poland

Acknowledgments

This book has been financially supported by several research and educational grants:

- Ministry of Science and Higher Education (MNiSW)—PMI–Projekty Ministra program, Grant No. NDS-II/SP/0411/2023/01. Project: "Advanced Medical Education in Extended Reality (edu-XR)"
  Period: December 18, 2023–December 18, 2026
- Foundation for the Development of the Education System (FRSE)—Erasmus+ program (Key Action 2), Grant No. 2023-1-PL01-KA220-HED-000159314. Project: "Interactive 3D Heart Anatomy Teaching Supported by Mixed Reality (iTeam 3D-MR)"
  Period: September 1, 2023–August 31, 2026
- National Science Centre (NCN)—OPUS program, Grant No. 24/55/B/NZ5/00494. Project: "Inverse Modelling of Repolarization Heterogeneity to Determine Arrhythmogenic Risk"
  Period: 2025–28
- National Science Centre (NCN)—Sonata Bis, Grant No. 2022/46/E/ST7/00008. Project: "Investigation of the Possibility of Development of Metaplasmonic Sensors for the Detection of Diabetes and NAFLD Biomarkers in the TDW/GLAD Technology"
  Period: 2023–28
- National Science Centre (NCN)—Preludium, Grant No. 2025/57/N/ST7/02502. Project: "Smart Breath: An AI-Based e-Nose Platform for Detection and Monitoring of Metabolic Syndrome Through Volatile Biomarkers"
  Period: 2026–28

The scientific knowledge and experience used in preparing this book were also developed within previously completed projects:

- National Centre for Research and Development (NCBR)—Lider program, Grant No. LIDER/17/0064/L-11/19/NCBR/2020. Project: "Holographic MedAssistant" finished
- Erasmus+ program, Grant No. 2020-1-PL01-KA203-HE-082077. Project: "Mixed Reality Supporting Advanced Medical Education" finished

CHAPTER

Introduction

1

Artur Rydosz[1,2], Klaudia Proniewska-van Dam[2,3] and Peter van Dam[2,4]

[1]*Biomarkers Analysis LAB, Institute of Electronics, AGH University of Kraków, Kraków, Poland,*
[2]*Functional and Virtual Medical 3D Imaging Laboratory, Department of Diagnostic Imaging, University Hospital in Kraków, Kraków, Poland,*
[3]*Center for Digital Medicine and Robotics, Jagiellonian University Medical College, Kraków, Poland,*
[4]*Department of Radiology, Jagiellonian University Medical College, Kraków, Poland*

1.1 Introduction

*"**The important thing is not to stop questioning. Curiosity has its own reason for existing**. One cannot help but be in awe when he contemplates the mysteries of eternity, of life, of the marvelous structure of reality. It is enough if one tries merely to comprehend a little of this mystery every day. Never lose a holy curiosity."*

—Albert Einstein

Personal Memoirs of William Miller, an editor, Life, May 2, 1955

Nowadays, the interdisciplinary approach for teaching and practicing medicine is a fact. It paves the way for hand-in-hand collaboration among specialists from diverse fields, including clinical teams, biomedical engineers, and computer scientists. Among many possible areas of cooperation, three-dimensional (3D) imaging is a key component of modern cardiovascular medicine. Probably never before have we been witnesses of technological progress so fast that not decades but single years are needed to change the clinical recommendation, and in the case of 3D imaging techniques, we can observe the enhanced clinical evidence that deeper cardiac structure and function can be modeled compared to the two-dimensional (2D) approaches. This technological advancement, the use of a new tool, requires a new education model for future specialists in cardiology, as well as for general practitioners who can apply these outcomes in their practices. An obvious question arises: what can 3D techniques add to the well-known clinical standard?

For example, functional imaging techniques enable real-time monitoring and analysis of dynamic physiological processes, thereby providing the clinical team with more information about the current state and potential progression of disorders. In cases of suspected coronary artery disease (CAD), 3D techniques may provide physiological information, reducing the need for unnecessary angiograms. A better understanding of heart anatomy and leaflet dynamics will be useful in all valvular

Functional and Virtual Medical 3D Imaging in Cardiovascular Medicine. DOI: https://doi.org/10.1016/B978-0-443-44570-5.00011-2

disorders, while tissue characterization plays a crucial role in the clinical evaluation of cardiomyopathy and myocarditis—all these elements can be obtained from 3D techniques. In addition, 3D echocardiography (3DE), a novel technique, has significantly improved volumetric and functional evaluation of cardiac chambers, overcoming the geometric assumptions of 2D imaging. Left atrial (LA) and right ventricular (RV) volumes and strain can be more accurately assessed via the 3DE method. Moreover, cardiac MRI (CMRI) can be supported by a four-dimensional (4D)-flow approach, which provides enhanced assessment of abnormal flows that may suggest heart failure in the coming weeks or months.

The benefits of the advanced method, such as advanced electrophysiological imaging versus conventional methods in cardiology, are well described in the next chapters, with special emphasis on immersive technologies such as extended reality (XR), virtual reality (VR), augmented reality (AR), and mixed reality (MR). Immersive technologies are revolutionizing how people interact with digital environments, and AI plays a key role in their development, making immersive experiences more intelligent, flexible, and user-oriented. Over the last few years, we have moved from static digital atlases and 2D cross-sectional reconstructions to 3D models, where patient-specific cases are the key. From a clinical perspective, these technologies can be applied in preoperative planning, medical education (including holographic representations), visualization of diagnostic data, and monitoring of clinical procedures, ultimately improving the safety and precision of therapeutic interventions. It is important to note that no single diagnostic technique fully describes the heart, which is characterized by complex geometry, dynamic function over time, and a strong interdependence between structure and electrophysiology. Therefore, by combining all possible 3D techniques (each technique has its own advantages), we can move from "what do we see?" to "what can we do?". Such an approach is, in fact, a game-changer for thinking about novel teaching processes and supporting clinicians in planning procedures covering areas of the cardiovascular system.

Chapter 2 presents that 3D imaging has become a core element of contemporary cardiovascular medicine, expanding clinical capabilities beyond traditional 2D techniques. Advanced modalities—including 3D echocardiography, cardiac CT, cardiovascular MRI with tissue characterization and 4D-flow, hybrid positron emission tomography (PET)/single-photon emission computed tomography (SPECT) imaging, patient-specific 3D printing, and immersive VR/AR visualization—support more accurate diagnosis, procedural planning, device selection, navigation, and personalized therapy. The chapter presents an evidence-based overview of technical principles, clinical applications, workflow integration, validation strategies, and regulatory considerations required for routine clinical adoption.

In electrophysiology, combining functional 3D datasets with immersive visualization enhances anatomical understanding, improves targeting of arrhythmogenic substrates, may reduce radiation exposure, and enables emerging low- or zero-fluoroscopy approaches, although outcomes depend on operator expertise and image-registration quality. Overall, the text highlights the transition of 3D imaging from a research innovation to an essential clinical tool across cardiology, outlining practical strategies for integrating advanced imaging into everyday medical practice.

Chapter 3 focuses on the physical and technical aspects of obtaining cardiac diagnostic data, processing, reconstructing, and presenting them in a clinically applicable way. It shows the heart as an organ that is a highly complex mechanical-electrical system, posing challenges for assessment, description, and, consequently, diagnostics. The evolution of diagnostic methods from 2D ultrasound projection to advanced multimodal spatial imaging based on dynamic 3D-CT is at the heart of treatment planning, and complex interventional procedures, and leads to the personalization of medicine. Examples of modern data use in clinical practice, their impact on the course of treatment, and patient monitoring in the subsequent period will be presented. The problem of optimizing data collection and processing processes, considering interpersonal cooperation, system integration, costs, and the need to improve personnel qualifications, will also not be overlooked.

Chapter 4 describes a rapidly developing field of modern medicine driven by interdisciplinary collaboration among clinicians, biomedical engineers, and computer scientists. It focuses on the integration of medical imaging, data analysis, artificial intelligence (AI), and immersive technologies (VR, AR, and MR) to create advanced 3D clinical tools that support preoperative planning, medical education, diagnostic visualization, and procedural monitoring, which may ultimately improve safety and precision in therapy; however, more clinical trials are needed to confirm this hypothesis.

It presents the full clinical and technological workflow, including digital imaging and communications in medicine (DICOM) data segmentation, multimodal imaging integration, software development, hospital system interoperability (picture archiving and communication system (PACS)/ radiology information system (RIS)/ hospital information system (HIS)), and data quality assessment. Special attention is given to functional imaging and spatiotemporal analysis, combining anatomical and physiological information. The chapter also discusses practical barriers to clinical implementation, such as workflow integration and interoperability challenges.

Drawing on the experience of the 3D-FM Laboratory in Kraków, the text outlines a system-oriented framework that transforms raw medical data into immersive 3D decision-support solutions, offering clinicians, engineers, and researchers structured guidance for adopting these technologies in routine practice and education.

Chapter 5 outlines the growing role of immersive and 3D technologies—including extended reality (XR), virtual reality (VR), mixed reality (MR), and advanced 3D imaging—in cardiovascular education. These tools enable clinicians and learners to visualize complex cardiac anatomy and pathology beyond traditional 2D imaging, supporting procedural planning, training, and clinical decision-making. The evolution of cardiac visualization has progressed from static atlases to interactive holographic and functional 3D models that enable dynamic interaction with anatomical and physiological data, including hemodynamic simulations.

Immersive technologies are also transforming medical education by improving spatial understanding, learner engagement, and procedural skill acquisition, while highlighting the need for structured curricula and validated evaluation methods. Platforms such as HoloAnatomy demonstrate how mixed-reality solutions can enhance anatomy teaching and clinical training.

In parallel, advanced imaging modalities (CT, MRI, and 3D echocardiography) and computational modeling approaches (computational fluid dynamics (CFD) and finite element analysis (FEA)) are becoming integrated into cardiovascular workflows, enabling patient-specific simulations for structural heart interventions. Effective implementation depends on interoperability standards such as DICOM and fast healthcare interoperability resources (FHIR), as well as integrated workflows combining segmentation, modeling, and immersive visualization. The chapter ultimately presents practical experience with functional and virtual 3D imaging, discussing both the opportunities and current challenges of adopting immersive technologies in routine cardiovascular education.

Chapter 6 explores the electrical and mechanical connections in heat: the electrical activation of the heart initiates mechanical contraction, followed by the active electrical recovery, which regulates mechanical relaxation. The propagation velocity in myocardial tissue, the pacemaker cell function, the sinus node, and the triangle of Koch, His-node, His-Purkinje system, and many other terms are explained from both perspectives giving the possibility to develop an electro-anatomical mapping system (EAM), where the propagation of the activation can be accurately measured; however, these EAM maps can only measure the local activation time on the endocardial or epicardial surface from which the 3D velocity has to be derived.

Quite interesting data are presented in Chapter 7, where the exhaled breath analysis (EBA) method is elaborated for heart failure monitoring. Nowadays, noninvasive methods are not only based on the physical analysis of signals, such as ECG/electrocardiography imaging (ECGI), but also on biomarker detection, including exhaled volatile organic compounds. Conventionally, blood/urine samples are collected to measure specific compounds, such as N-terminal pro-B-type natriuretic peptide (NT-proBNP), in heart failure. However, EBA has the potential to serve as a supplementary tool in daily clinical practice; it can be used in combination with 3D imaging and functional tests, which are discussed in detail in this chapter.

It must be understood that not all centers have advanced equipment such as CT-FFR or MRI/CMRI tools, as well as well-qualified medical staff to both interpret the data and conduct medical procedures, making use of the latest advances in this field. Additionally, despite growing interest and increasing availability of 3D functional imaging, its clinical translation in cardiology remains constrained by interoperability and integration challenges within existing clinical workflows. The General Data Protection Regulation (GDPR) must be applied to ensure effective management of clinical data, including structuring, anonymization, and compliance with data protection requirements, among other measures, especially when AI models are planned for use. Therefore, in Chapter 8, the approaches for medical data analysis support are presented. In modern clinical practice, the diagnostic process is no longer a solitary endeavor between a physician and a patient. Due to several factors, such as a lower number of clinicians per citizen, the need to work in remote areas, an aging population, and an increased number of cardiovascular disease (CVD) episodes among the younger population, the Digital Consultant is playing a key role. Over the decades of thinking about this solution as a science-fiction dream, we progressed to Computer-Aided Detection (CAiD) and Computer-Aided Diagnosis (CADx) methods. However, the bridge between a new computational algorithm and a life-saving clinical decision is built not on code but on the rigorous discipline of biostatistics, where Clinical Decision Support (CDS) and Software as Medical Device (SaMD) methods are continually refined and developed.

Overall, one of the major goals of this book is to provide a comprehensive overview of novel developments in cardiology from a practical perspective. This book presents key points, recommendations, and checklists for those who want to embark on a journey into the 3D world of cardiology.

CHAPTER 2

Heart and cardiovascular system: advanced 3D imaging techniques in cardiology: applications, clinical integration, and the future of cardiovascular diagnostics

William Uribe[1,2], Klaudia Proniewska-van Dam[2,3] and Peter van Dam[3,4]

[1]Department of Cardiology and Electrophysiology, CES University, Medellin, Colombia,
[2]Center for Digital Medicine and Robotics, Jagiellonian University Medical College, Kraków, Poland,
[3]Functional and Virtual Medical 3D Imaging Laboratory, Department of Diagnostic Imaging, University Hospital in Kraków, Kraków, Poland,
[4]Department of Radiology, Jagiellonian University Medical College, Kraków, Poland

2.1 Introduction

Cardiovascular imaging is at the forefront of diagnostic and therapeutic strategies in clinical cardiology. With the rapid evolution of technology, three-dimensional (3D) imaging techniques have become essential tools, offering more profound insights into cardiac structure and function beyond traditional bi-dimensional (2D) modalities. These advanced methods, spanning echocardiography, cardiac magnetic resonance (CMR), computed tomography (CT), and nuclear techniques, enable more accurate assessments, allow preprocedural planning to avoid surprises during the intervention, and enhance patient outcomes.

3D imaging has evolved from a research tool to a clinical mainstay across cardiology subspecialties. In common daily medical practice, decisions about diagnosis, treatment, and monitoring increasingly rely on imaging. 3D imaging gives more precise information on volumes, valve anatomy, flow dynamics, and tissue characterization. Incorporating this new imaging into common practice enables better risk stratification, better selection of patients for invasive procedures,

Functional and Virtual Medical 3D Imaging in Cardiovascular Medicine. DOI: https://doi.org/10.1016/B978-0-443-44570-5.00007-0

and often reduces unnecessary tests. This chapter provides a comprehensive and updated review of the available 3D imaging modalities, supporting evidence, clinical indications, and potential pitfalls, while highlighting how the integration of advanced 3D imaging enhances and complements traditional 2D-based approaches in routine clinical practice [1–4].

In countries within the Global South, such as Colombia, the adoption of advanced imaging technologies for integration into cardiac electrophysiology has progressed steadily, though implementation varies across centers. National initiatives, including the Colombian Guidelines of Cardiovascular Electrophysiology, have emphasized the role of 3D mapping systems and multimodal image integration in improving procedural precision and patient safety (Fig. 2.1). These guidelines provide structured recommendations for the incorporation of CT, MRI, and multimodality fusion into arrhythmia management, particularly in complex ablations. As highlighted in an editorial by one of the authors of this chapter, the development of cardiovascular electrophysiology in Colombia requires the systematic incorporation of such innovations to reduce radiation exposure, optimize procedural workflows, and align practice with international standards [5,6].

This chapter explores the clinical integration of advanced 3D imaging in cardiology, focusing on how these complement and extend classical imaging

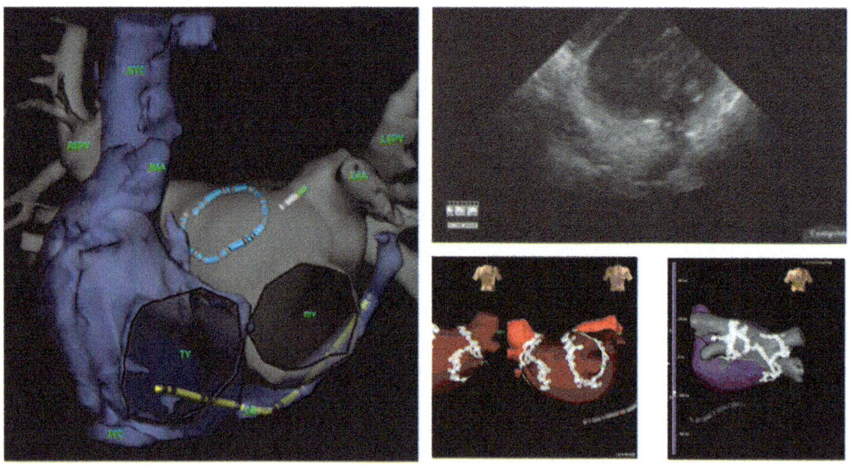

FIGURE 2.1 LEFT:

CCT segmentation and reconstruction of the right and left atrium with principal veins. Right superior: Intracardiac echocardiography of the left atrium, showing a circular catheter on the left inferior pulmonary vein. Right inferior: 3D electroanatomic mapping of the left atrium integrated with EnSite system software. White dots showing points of radiofrequency application around the pulmonary veins antrum and posterior wall.

approaches. The following sections outline the key imaging modalities, acquisition principles, quantitative analysis, and clinical integration strategies necessary for physicians to implement advanced 3D imaging in their clinical practice.

2.2 Imaging modalities and when to use them

The most common clinical questions in cardiology, along with the suggested optimal imaging modalities and the added value of 3D techniques [7–10], are presented in Table 2.1.

The most common clinical questions in electrophysiology, along with the suggested optimal imaging modalities and the added value of 3D techniques and their integration [11–14], are presented in Table 2.2.

2.2.1 Real-time 3D echocardiography

Three-dimensional echocardiography (3DE) has significantly improved volumetric and functional evaluation of cardiac chambers, overcoming the geometric assumptions of 2D imaging. For instance, 3DE provides more accurate assessments of left atrial (LA) and right ventricular (RV) volumes and strain, both of which hold prognostic significance across a wide spectrum of cardiac diseases.

Clinical integration is most evident in valvular disease: 3DE facilitates precise visualization of mitral and tricuspid valve anatomy during interventions like transcatheter edge-to-edge repair, providing real-time intraprocedural guidance. Its application in RV analysis also contributes to prognostic assessment in pulmonary hypertension and heart failure. Still, comparative studies with CMR

Table 2.1 Modality comparison: best modalities for clinical questions in cardiology and what are the advantages of incorporating 3D.

Clinical question	Best modality(ies)	What the 3D adds
Suspected coronary artery disease (CAD), stable chest pain	CTA + CT-FFR	Adds physiologic information; reduces unneeded angiograms
Valvular disease (MR, AS, TR)	3D echocardiography + CT (for TAVR)	Better anatomy, leaflet dynamics; more accurate sizing
Cardiomyopathy, myocarditis	CMR (LGE, T1/T2 mapping)	Tissue characterization: scar vs. inflammation
Patient follow-up/serial assessment	3D echocardiography and MRI	More reproducible measures with 3D vs. 2D

Abbreviations: *AS*, aortic stenosis; *CTA*, computed tomography angiography; *CT*, computed tomography; *EP*, electrophysiology; *FFR*, fractional flow reserve; *MR*, mitral regurgitation; *TR*, tricuspid regurgitation.

Table 2.2 Modality comparison for key clinical questions in electrophysiology.

Clinical question	Preferred imaging modality (2D/ standard)	Best advanced/ 3D modality	Key advantages of 3D integration
Atrial fibrillation (AF) ablation planning	Transesophageal echocardiography (TEE), fluoroscopy	CT or MR angiography fused with 3D electroanatomic mapping (EAM)	Precise pulmonary vein (PV) anatomy; reduced fluoroscopy; improved lesion set coverage; single-procedure success ≈75%–85 %
Left atrial appendage (LAA) thrombus exclusion/ structural evaluation	TEE	3D TEE or cardiac CT	Better ostial sizing and morphology for occlusion devices; improved procedural guidance
Scar characterization in VT or atypical atrial tachycardia	Conventional MRI (late-gadolinium enhancement)	High-resolution 3D LGE-CMR fused with EAM	Identifies critical isthmuses; reduces VT recurrence (32% vs. ~48% at 1 year in multicenter data)
Accessory pathway/ pediatric ablations	Fluoroscopy	3D EAM ± intracardiac echocardiography (ICE)	"Zero-fluoro" workflow lowers radiation by >70 %
Complex congenital heart disease EP mapping	2D echocardiography, fluoroscopy	3D rotational angiography or CT/ MR with EAM	Depicts complex anatomy, aids catheter navigation, decreases procedure time
CRT or His-bundle lead placement	Fluoroscopy, venography	3D CT venography or fusion with EAM	Accurate coronary-sinus branch mapping; improved success of target lead positioning
Real-time lesion visualization	Intracardiac echocardiography (ICE)	3D ICE or MRI (experimental)	Continuous lesion monitoring; assessment of edema and gap formation

Abbreviations: AF, Atrial fibrillation; *CRT*, cardiac resynchronization therapy; *EAM*, electroanatomic mapping; *ICE*, intracardiac echocardiography; *LAA*, left atrial appendage; *LGE-CMR*, late gadolinium enhancement cardiac MRI; *PV*, pulmonary vein; *VT*, ventricular tachycardia.

are needed to solidify normative values and resolve discrepancies, especially in strain measurements [15–17].

Acquisition modes include full-volume gated multibeat, real-time single-beat, and color Doppler 3D volumes for valve quantification. Key technical parameters:

voxel anisotropy, temporal resolution, sector width, and depth. Quantitative applications span chamber volumetry, vena contracta area, regurgitant volume (3D Proximal Isovelocity Surface Area (PISA)), and leaflet geometry for mitral/tricuspid repair. Limitations include acoustic shadowing, limited field of view, and operator dependence.

- Clinical utility: accurate chamber volumes, ejection fraction, valve morphology, and more reliable measures of regurgitant orifice than 2D echo.
- Evidence: a meta-analysis of real-time (RT)-3DE versus CMR shows high agreement for ejection fraction, ventricular mass, though 3DE often underestimates volumes, especially end-diastolic volume/end-systolic volume (EDV/ESV) in older patients or those with disease [18].
- Practical limitations: image quality in obese patients, acoustic windows; learning curve for acquisition and interpretation.

2.2.2 Computed tomography angiography + coronary centerline-based fractional flow reserve estimation for coronary disease

For 3D reconstruction, electrocardiogram (ECG)-gated acquisition with isotropic voxels is essential. Postprocessing employs multiplanar reformation, volume rendering, and centerline extraction. Advanced tools include plaque characterization, coronary centerline-based FFR estimation (CT-FFR), and virtual stent planning. Dose modulation and iterative reconstruction reduce radiation exposure.

- What it offers: noninvasive anatomic imaging + functional/ischemia prediction without stress tests or invasive angiography.
- Key studies: The TARGET (Targeted Ablation of Rotors for the Treatment of Persistent Atrial Fibrillation) randomized trial demonstrated that use of on-site CT-FFR reduced the number of invasive angiograms that showed no obstructive disease [19].
- Cost/accessibility: Must consider radiation, contrast risks, and availability of computational or artificial intelligence (AI) FFR tools. Some trials show good cost-effectiveness in selected populations.

2.2.3 Cardiac MRI and 4D flow

CMR offers unparalleled tissue characterization and volumetric assessment, with 3D cine sequences and strain analysis now part of routine evaluation. A notable advancement includes the use of CMR-derived intraventricular pressure gradients, which offer new functional insights, particularly post-myocardial infarction, and correlate with long-term outcomes [19–21].

In the context of cardiomyopathies and heart failure, 3D late gadolinium enhancement (LGE) mapping and radiomics techniques enable quantification of myocardial scar heterogeneity and stratification of sudden cardiac death risk.

These features enhance classical LGE analysis by incorporating shape, distribution, and texture.

Techniques include ECG-gated 3D cine balanced steady-state free precession, isotropic LGE, T1/T2 mapping, and 4D flow MRI for volumetric velocity fields. Strengths include gold-standard ventricular volumes, tissue characterization, and flow quantification. Key technical aspects: compressed sensing, parallel imaging, and motion compensation.

- Clinical strengths: scar imaging LGE, accurate volumes/mass, quantification of diffuse fibrosis (T1/T2 mapping), and ability to assess abnormal flows.
- Evidence: for many core measures (volumes, mass, EF), CMR remains the reference standard; 3DE correlates well but generally underestimates volumes.

2.2.4 Hybrid/nuclear imaging and 3D modeling

Positron emission tomography (PET) and Single Photon Emission Computed Tomography (SPECT) continue to benefit from hybrid 3D imaging integration. Recent developments include 3D myocardial flow reserve mapping and combined viability/perfusion assessments, which add critical prognostic value in ischemic cardiomyopathy.

3D PET imaging of coronary microcalcification and inflammation using 18F-sodium fluoride has shown strong associations with myocardial infarction risk, offering a molecular-level complement to anatomical CT. These innovations are particularly valuable when classical tests yield inconclusive results.

Hybrid PET/CT and SPECT/CT combine 3D perfusion imaging with coronary anatomy, supporting ischemia evaluation and viability assessment (Fig. 2.2). Digital PET enhances resolution for 3D perfusion and metabolic mapping [22,23].

- Useful when perfusion is needed (e.g., for ischemia, viability). Hybrid PET/CT or SPECT/CT gives 3D maps of perfusion plus anatomy (Fig. 2.3).
- 3D printing or virtual modeling helps in complex valve disease, structural heart interventions, and may help in patient education.

2.2.5 Cardiac computed tomography: from anatomy to AI-powered risk prediction

Coronary CT angiography (CCTA) remains the modality of choice for noninvasive coronary artery visualization. Recent 3D innovations include photon-counting CT, which enhances spatial resolution while reducing radiation exposure and the need for follow-up imaging. The integration of AI with 3D CT imaging has transformed risk prediction. AI-powered quantitative coronary plaque analysis and virtual fractional flow reserve computation enhance diagnostic accuracy and therapeutic planning, particularly in stable ischemic

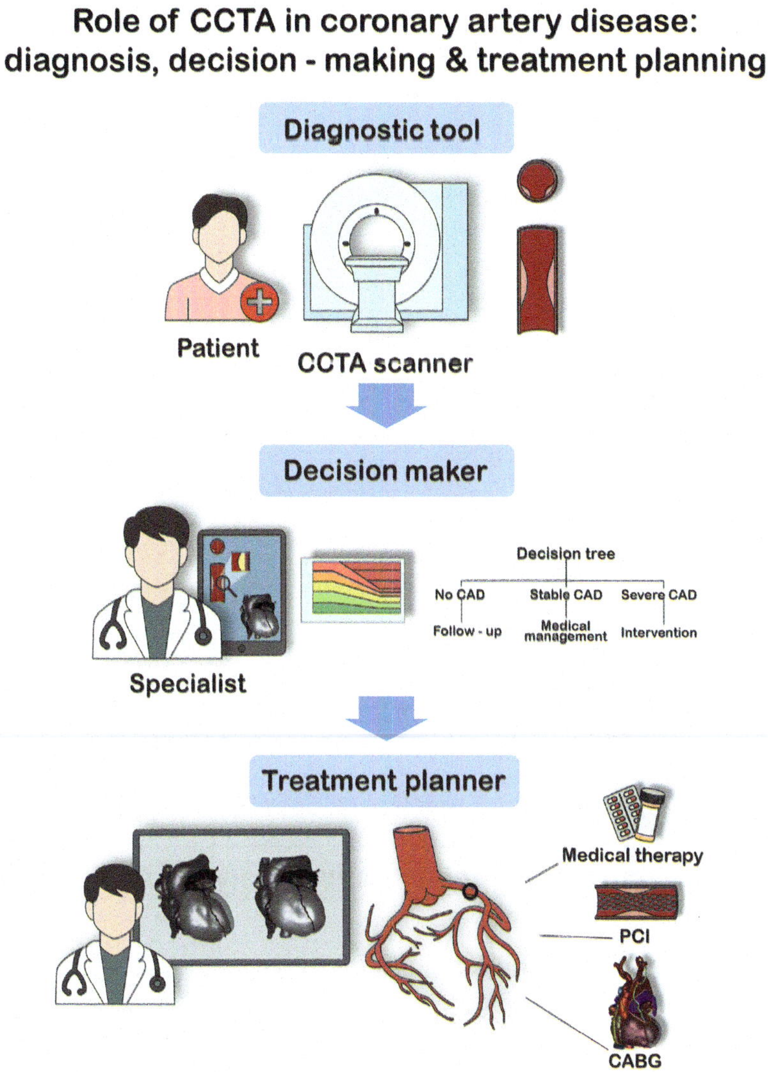

FIGURE 2.2

The role of CCTA in coronary artery disease: a diagnostic tool, decision-maker, and treatment planner. *CCTA*, coronary computed tomographic angiography; *CAD*, coronary artery disease; *PCI*, percutaneous coronary intervention; *CABG*, coronary artery bypass graft.

heart disease (Fig. 2.4). Additionally, 3D CT datasets now allow for fully automated assessment of left ventricular and atrial function, increasingly matching or surpassing manual evaluation in speed and reliability [24,25].

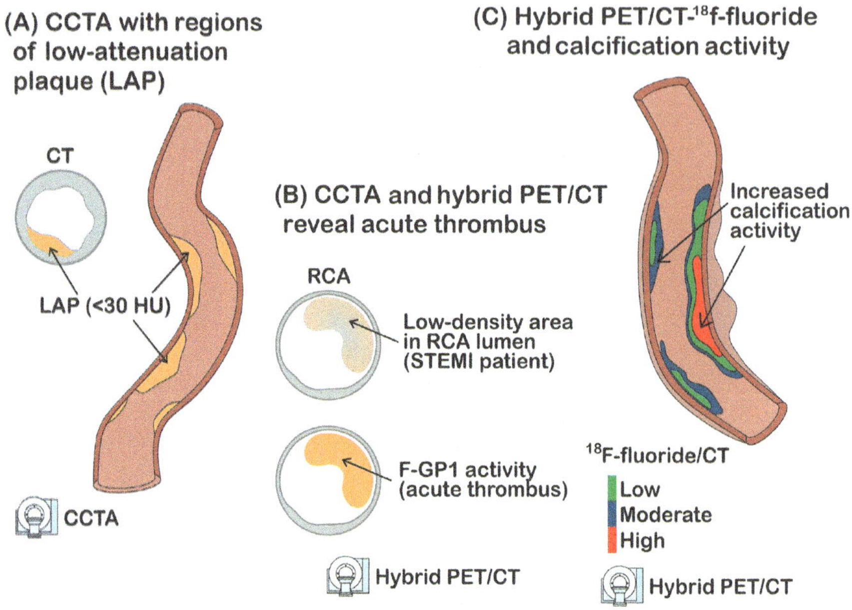

FIGURE 2.3

Advanced assessments of plaque type, plaque thrombosis, and disease activity with CT and hybrid PET/CT. (A) CCTA with regions of LAP (<30 HU, orange areas). (B) CCTA showing a low-density area in the lumen of RCA in a patient with inferior STEMI. Hybrid PET/CT images demonstrate 18F-GP1 activity as acute thrombus in this region (yellow area). (C) Hybrid PET/CT image after administration of 18f-fluoride. Colored areas represent regions of increased calcification activity.

2.3 Image processing, quantitative analysis, and artificial intelligence [26–29]

2.3.1 Segmentation and landmarking

Automated and semiautomated segmentation (active contours, level sets, graph-cuts, convolutional neural networks) enables reproducible extraction of cardiac structures. Validation should include the Dice similarity coefficient, or Dice coefficient, a statistical method of spatial overlap, Hausdorff distance (a measure of the strength and direction of a linear relationship between two quantitative variables), and correlation with invasive measurements.

2.3.2 Machine learning and deep learning

Applications include accelerated reconstruction, automated quality control, plaque classification, scar detection, and outcome prediction using 3D biomarkers. Clinicians must evaluate AI tools for intended use and potential bias.

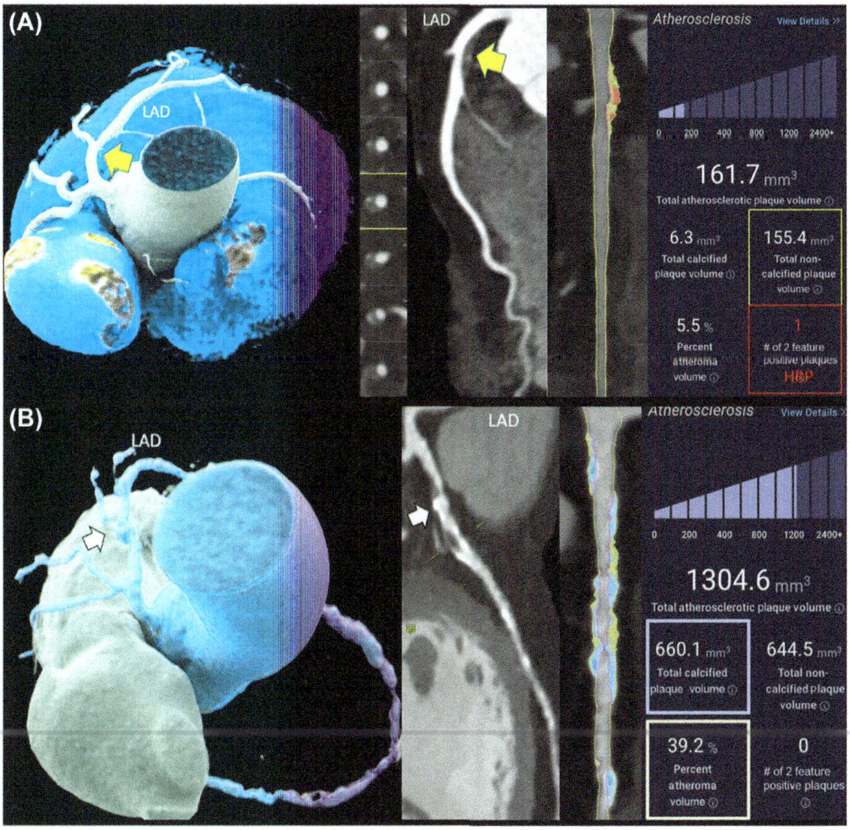

FIGURE 2.4

Artificial intelligence–based quantitative computed tomography (AI-QCT) plaque analysis. Case examples: (A) A 55-year-old woman with atypical chest pain and two cardiovascular risk factors (CVRFs): arterial hypertension, dyslipidemia, obesity, and hyperuricemia. Coronary computed tomography angiography (CTA) showed a noncalcified plaque (NCP) in the proximal left-anterior descending (LAD) coronary artery (yellow arrow) with <50% diameter stenosis. Right: Artificial intelligence–based quantitative computed tomography (AI-QCT) plaque analysis showed an NCP volume of 155.4 mm^3 (yellow box) and high-risk-plaque (HRP; red box) features. Note that color overlay (mid) facilitates the delineation of NCP and HRP. (B) A 65-year-old woman with atypical chest pain, dyspnea, and a high coronary risk profile (5 CVRFs), smoking, arterial hypertension, positive family history, dyslipidemia, diabetes, and high lipoprotein A (89.1 ng/dL), who experienced a major adverse cardiovascular event and underwent percutaneous coronary intervention (PCI) (LAD prox/mid/distal and right coronary artery [RCA]). Coronary CTA: white arrow pointing at calcified plaque in the proximal LAD. Mid, curved multiplanar reformation. Right: AI-QCT plaque analysis revealed high total plaque volume (stage 3) with high calcified, high non-calcified plaque volume, and high percentage atheroma volume (39.2%).

Reprinted from Feuchtner GM, Lacaita PG, Bax JJ, Rodriguez F, et al. AI-quantitative CT coronary plaque features associate with a higher relative risk in women: CONFIRM2 registry. Circulation: Cardiovascular Imaging 2025. CC BY 4.0. @2025, Wolters Kluwer Health. doi: 10.1161/CIRCIMAGING.125.018235 PubMed: PMID 40162910.

2.3.3 Computational modeling and virtual physiology

CT-FFR combines 3D lumen geometry with physiologic modeling for lesion-specific ischemia assessment. Computational fluid dynamics enables the simulation of shunts, valve interventions, and device–blood interactions.

2.4 Clinical integration: merging classic and advanced views

A pivotal aspect of 3D imaging's role in clinical practice is not replacement, but integration. For instance:

In valvular intervention, 3DE and 3D CMR guide anatomical assessment, while classical 2D echo remains essential for real-time function monitoring.

In cardiomyopathies, 3D LGE or strain patterns complement echocardiographic findings and ECG interpretations, particularly in risk stratification.

In coronary artery disease, AI-augmented 3D CT and PET imaging now assist in guiding interventions when classical angiography is ambiguous or inconclusive.

In electrophysiology: Electrophysiology has traditionally relied on fluoroscopy, classic electroanatomic mapping, and activation/pace mapping to locate arrhythmogenic substrates. However, merging these classic views with advanced imaging modalities, CT/MR integrations, image guidance systems, mixed or augmented reality, and AI-assisted overlays, offers enhanced anatomical and functional context. Clinical trials have shown that integrating CT into electroanatomic mapping can improve procedural success and safety in atrial fibrillation ablation, while modules like fluoroscopy integration reduce radiation exposure substantially without compromising outcomes. Implementing this merger requires attention to imaging/rhythm alignment, operator training, and workflow adaptation. For general electrophysiologists, the benefit lies in greater precision, lower risk, and possibly shorter learning curves when tackling complex anatomies.

In EP, "classic views" refer to traditional techniques and imaging modalities: fluoroscopy, standard electroanatomic mapping (EAM), catheter manipulation and activation mapping, and less image-guided catheter ablations. "Advanced views" are those that incorporate newer imaging or guidance technologies, CT/MR image integration, 3D imaging, digital twins, augmented or mixed reality, AI guidance, etc. Clinical integration is about combining the strengths of both to improve outcomes, safety, and procedural efficiency.

Why is this integration important?

The answers to this question can be easily understood in Table 2.3.

- Accessory pathway ablation + FIM: In a cohort of 131 patients, use of a fluoroscopy integration module reduced fluoroscopy time from ~7.4 to ~2.8 minutes and dose area product from 12.47 to 5.8 μGy m^2 without reducing success [30].

Table 2.3 Integration benefits in electrophysiology. See below the evidence on clinical uses, benefits, and pitfalls.

Benefit	Description/examples
Reduced radiation exposure	For example, combining a fluoroscopy integration module (FIM) with fluoroscopy in accessory pathway ablation reduced radiation exposure substantially.
Better anatomical accuracy, procedural safety	Integrating CT imaging into EAM (CT-EAM integration) helps better visualize pulmonary veins, left atrial or posterior anatomy, reducing complications like pulmonary vein stenosis.
Consistent outcomes over operator experience	Even less experienced operators benefit when guided by advanced imaging overlays or image integration; variation in success tends to decrease. For example, the FIM reduced fluoroscopy time regardless of operator experience.
Improved decision-making and visualization	Mixed/augmented reality systems overlay imaging onto the live procedural field, giving physicians real-time spatial orientation. This helps in navigation, catheter placement, and avoidance of injury to critical structures.

Abbreviations: *EAM*, Electroanatomic mapping; *FIM*, fluoroscopy integration/imaging management module.

- CT + EAM in AF ablation: The 2007 et al. study compared "Carto MERGE" (CT-EAM image integration) with EAM alone for AF ablation; success rates were higher with image integration (85.1% vs. 67.9%) and safety was better (less pulmonary vein stenosis), although procedure/fluoroscopy times did not change [31].
- Image integration accuracy: A study strongly suggested that MR imaging of the left atrium integrated into EAM has similar registration error irrespective of whether patients were in AF or sinus rhythm at imaging or ablation. This supports the usability of advanced imaging even when rhythm differs [11].

A very comprehensive comparative workflow diagram of classic versus advanced electrophysiology imaging is shown in Fig. 2.5.

An integrated 3D navigation scene at the EP lab, suitable for beginners in the topic, can be seen in Fig. 2.2. An overview of an integrated 3D navigation environment in the electrophysiology lab (laboratory) is shown in Fig. 2.6. The routine use of fluoroscopy in interventional EP practice is steadily declining, largely due to the availability of advanced technologies such as 3-dimensional (3D) electroanatomic mapping, intracardiac echocardiography (ICE), rotational CT angiography combined with real-time CT or MRI fusion, and the more recent technology of mixed reality incorporation, during clinical procedures in state-of-the-art specialized laboratories.

How to implement integration in everyday EP practice [32–34]

Here are steps and considerations for integrating classic and advanced views:

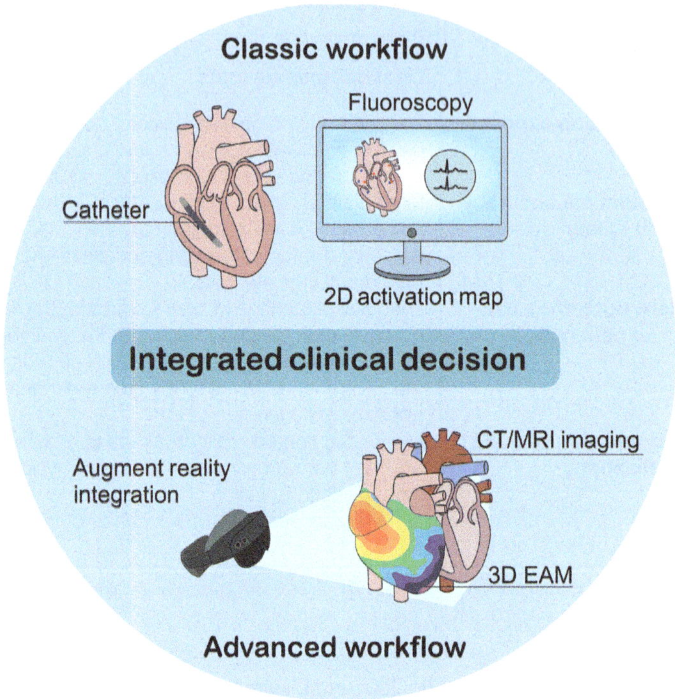

FIGURE 2.5 Classic versus advanced electrophysiology lab imaging workflow.

Superior image (classic): Catheter icon inside a heart silhouette. Fluoroscopy monitor showing a simple 2D outline activation map with a few electrode dots. Inferior image (Advanced): 3D heart rendering with color-coded electroanatomic map (EAM), overlaid CT/MR anatomy, mixed-reality headset icon, or augmented reality (AR). Center labeled: integrated clinical decision.

1. Preprocedural imaging

- Acquire high-quality CT or MRI (or both) before the procedure when anatomy is likely complex (e.g. in redo AF ablations, left atrial anatomy, congenital variants).
- Ensure rhythm and patient state (hydration, heart rate) are reasonably stable so imaging is usable.

2. Image registration/integration

- Use software systems that allow overlay of CT/MR data on electroanatomic mapping (EAM) systems. Validate registration accuracy during the procedure (some "landmark checks").
- Be aware of possible error sources: patient motion, rhythm differences, changes in anatomy due to previous ablations or remodeling.

3. Procedural guidance using mixed/augmented reality or real-time overlay

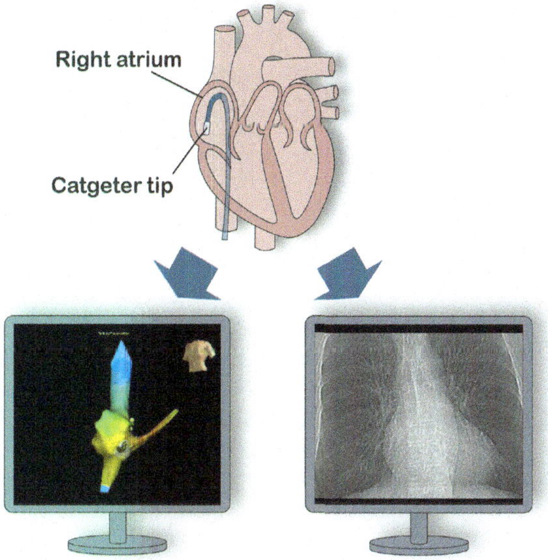

FIGURE 2.6 Fusion of classic and advanced views during ablation.

A perspective drawing of an EP lab where a catheter is shown navigating inside the heart → background monitors: one monitor displays a fluoroscopy image, and an adjacent monitor displays a 3D electroanatomic map fused with preacquired CT/MRI. Thin linking lines illustrate real-time synchronization of catheter location across both views.

Figure made with BioRender AI.

- Use mapping systems that display imaging in 3D alongside live electrograms/activation maps.
- If available, AR/MR headsets or head-mounted displays may allow viewing both the real instrument and its virtual location in anatomy (Fig. 2.7).

4. Operator training and workflow adjustments

- Operators familiar with classic methods need to be trained in interpreting integrated images.
- Time needs to be allocated for registration and validation steps. There may be minor procedural time overhead initially.

5. Quality tracking and feedback

- Track outcomes (success, recurrence, complications) and compare with prior cases where advanced imaging was/was not, used.
- Monitor radiation exposure, procedural time, mapping errors, etc., to ensure that the added complexity yields benefit.

Patient-centric outcome pathway [33]

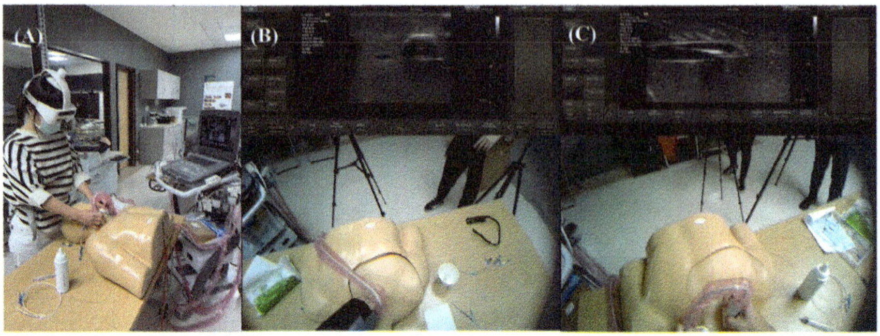

FIGURE 2.7 Practical operation.

(A) Trainee actively engaged in the process of executing Augmented Reality-Ultra Sound (AR-US). (B) and (C) Visual perspectives observed by the trainee using AR-US.

Reprinted from Liao, SC, Shao, SC, Gao, SY. et al. Augmented reality visualization for ultrasound-guided interventions: a pilot randomized crossover trial to assess trainee performance and cognitive load. BMC Medical Education 2024; 24: https://doi.org/10.1186/s12909-024-05998-8. 2024 Sep 27. CC BY 4.0, © 2024 The Author(s).

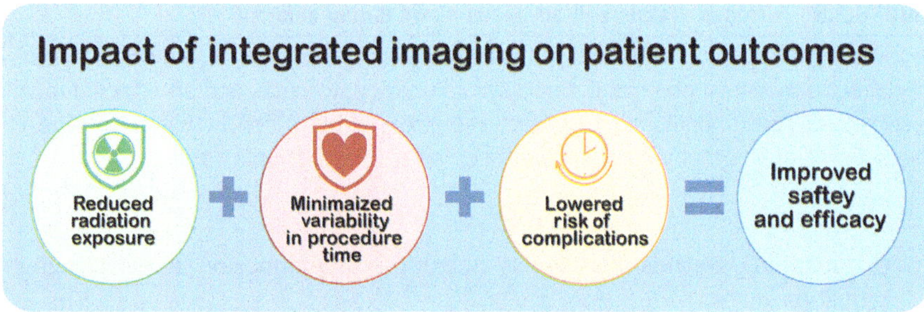

FIGURE 2.8 Impact of integrated imaging on patient outcomes.

Summatory leads to a central "Improved Safety and Efficacy" hub.

The primary purpose of adopting new technology in daily practice is to improve patient outcomes. Fig. 2.8 presents a circular flowchart showing how an integrated approach can reduce radiation exposure, minimize variability in procedure time, and lower the risk of complications.

Image integration: limitations and evidence gaps in electrophysiology

- No guaranteed improvement in all outcome metrics. Some randomized studies showed no significant difference in long-term success when image integration was added (e.g., in first AF ablations) [31,32].
- Registration error.

Differences in heart rhythm between imaging and ablation, patient motion (respiration, blood volume changes), or prior scar can degrade alignment [35].

- Procedure time, cost, resource requirements. Advanced imaging and registration may require extra software/hardware, trained personnel, and sometimes longer preparation time [36].
- Availability and access issues.

In many settings, access to high-resolution CT/MR, AR/MR equipment, or advanced mapping systems may be limited [37].

Other limitations include the heterogeneity of trials: randomized trials and registries differ in patient populations (paroxysmal vs persistent AF, first-time vs. redo procedures), imaging protocols, and EAM systems, making the results highly context-dependent. Registration error and physiologic mismatch: differences in atrial volume/geometry between preprocedural imaging and intraprocedural anatomy (due to rhythm, loading conditions, catheters) create registration errors that can mislead ablation if not recognized. Limited randomized outcome data in VT: robust randomized trials proving large outcome benefits (hard endpoints) for routine image-guided VT ablation are still limited; most data are matched cohorts and single-center series. Resource and training requirements: successful integration needs skilled imaging technologists, data-processing workflows, and regular multidisciplinary practice to avoid adding time without benefit [38–40].

Integration into practice: workflow for physicians

The above integrative approach ensures that clinicians make the most of both traditional experience and technological innovation; workflow is as follows:

1. Define the clinical question, for example, "Does this patient have ischemia? Is valve repair needed? What is the risk of arrhythmia?"
2. Choose the modality based on the question, availability, patient factors (age, comorbidities, renal function, contrast allergy).
3. Obtain high-quality images; work with the imaging center to ensure protocols (slice thickness, gating, contrast) are optimized.
4. Analyze and interpret, ensure measurements are reproducible, pay attention to known underestimation or bias (e.g., 3DE vs. MRI).
5. Decide on treatment; use information to guide medical therapy, invasive angiography, surgical referral, or device therapy.
6. Track outcomes and feedback, compare imaging predictions to actual findings and interventions to improve practice.

2.5 Clinical applications [30,31,33,37,40]

2.5.1 Coronary artery disease

3D CTA with CT-FFR offers a combined anatomic and functional assessment, reducing unnecessary invasive angiography. Hybrid PET/CT or PET/MR integrates perfusion defects with precise 3D localization.

2.5.2 Valvular heart disease

3DE quantifies leaflet tethering and annular dynamics, guiding transcatheter and surgical interventions. CT remains the standard for TAVR planning, while fusion imaging enhances intraprocedural accuracy.

2.5.3 Congenital heart disease

3D CMR and CT reconstructions, often converted to printed models, improve surgical and catheter-based intervention planning.

2.5.4 Heart failure and cardiomyopathies

Volumetric scar quantification and tissue characterization predict arrhythmic risk and guide device therapy. 4D flow evaluates abnormal flows and energy loss [21].

2.5.5 Electrophysiology and arrhythmia management

Image-based anatomical shells fused with intracardiac mapping refine ablation strategies, especially for scar-related VT and in complex arrhythmias such as atrial fibrillation/atrial flutter and atrial tachycardia. Augmented reality visualization may reduce fluoroscopy time, inclusive of becoming fluoro-less procedures [41].

What evidence shows [30–32,35,39,42]

Coronary disease: Use of CT-FFR has been shown to reduce unnecessary invasive angiography, with no increase in adverse events. TARGET trial data support this. Valvular Heart Disease: 3DE and CT improve measurement of the annulus, leaflet geometry; help size prostheses accurately in TAVR to reduce complications. Cardiomyopathy/myocarditis: CMR with LGE and mapping helps differentiate causes and predict outcomes such as risk of arrhythmia or need for ICD.

Complex arrhythmias: Image integration can improve procedural precision and reduce radiation. Multiple observational series and contemporary device/module studies demonstrate that integrating preprocedural CT/MR into electroanatomic mapping or using fluoroscopy integration modules reduces fluoroscopy time and dose without compromising acute success. This supports safer workflows, particularly for operators adopting low-fluoroscopy or fluoro-less techniques.

For atrial fibrillation ablation, CT/MR integration has produced mixed but clinically relevant results. Randomized and prospective studies show that CT integration can improve anatomic understanding and sometimes procedural outcomes (reduction in recurrences in some series), but effects are heterogeneous; benefits are most evident in complex anatomies, redo procedures, or when anatomic detail changes management. Registration accuracy and workflow matter.

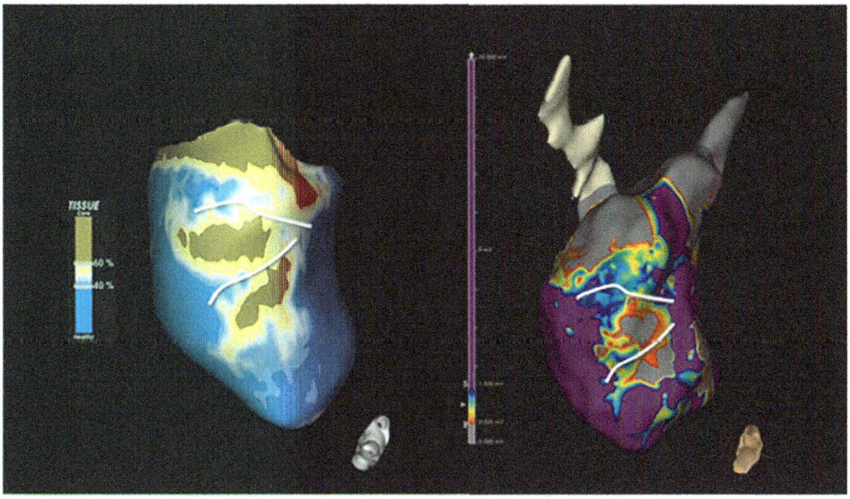

FIGURE 2.9 Correlation of the 3D-VI reconstruction from LGE-MRI with EAM.

Example of correlation of the 3D-VI reconstruction from LGE-MRI with electroanatomical voltage map. From left to right: LGE-MRI map (color-coding: blue: healthy tissue, yellow: border-zone, and red: core scar) with two clear conducting channels (white line) and an electroanatomical high-density voltage map showing septal scar (color-coding: purple: healthy tissue, red: border zone, and gray: core scar), with conducting channels at the same region as compared to the LGE-MRI.

Adapted from Roca-Luque I, Mont-Girbau L. Cardiac magnetic resonance for ventricular tachycardia ablation and risk stratification. Frontiers in Cardiovascular Medicine 2022;8:797864. doi: 10.3389/fcvm.2021.797864.

In scar-related ventricular tachycardia (VT), functional 3D substrate imaging helps localize targets. Recent work using late-enhancement CT or CMR-derived scar/critical-conduction-channel maps fused with EAM (Fig. 2.9) shows improved substrate localization and may improve acute/long-term VT control in selected cohorts (nonschemic and ischemic cardiomyopathy), particularly when combined with standardized transfer workflows. Contemporary single-center and matched-cohort studies report improved targeting and procedural metrics versus conventional mapping alone.

Virtual and augmented reality (VR/AR) are maturing into clinically cleared tools. Reviews and recent regulatory updates report growing use of VR/AR for preprocedural planning, intraprocedural guidance, and education. Several commercial AR/VR products for cardiovascular procedures have achieved regulatory clearance and early clinical adoption, and the literature supports improved spatial orientation and team communication (evidence base: early-phase trials, feasibility, and pilot studies).

Validation against ground truth and histology is improving confidence in imaging biomarkers. Studies that compare CMR/CT scar maps with histology and high-resolution electroanatomic mapping show reasonable spatial agreement and improved substrate identification, strengthening the biological plausibility of image-guided ablation targets, but registration errors and mapping resolution remain limiting factors. Serial monitoring: 3DE offers easier and faster follow-ups; MRI is preferred when high precision is needed or in complex patients.

2.6 Technical considerations and quality assurance

Standardized acquisition protocols, structured Digital Imaging and Communications in Medicine reporting, and multicenter reproducibility studies are critical for high-quality 3D imaging. Continuous validation with phantom and clinical outcome studies is recommended [43].

2.7 Integration into clinical practice

1. Triage: define the clinical question.
2. Modality selection: choose the most appropriate imaging modality.
3. Acquisition and quality check: perform on-table verification.
4. Processing and analysis: use validated software with human oversight.
5. Multidisciplinary review: involve the structural heart team and other stakeholders.
6. Documentation: track predictions versus outcomes for quality improvement.

2.7.1 Clinical integration in electrophysiology—evidence summary

The current data on functional and virtual 3D imaging in electrophysiology for a cardiologist:

The clinical adoption of advanced 3D imaging in electrophysiology has evolved from an adjunct to fluoroscopy into a cornerstone of modern ablation strategy. Integration of electroanatomic mapping (EAM) with computed tomography (CT) and cardiac magnetic resonance (CMR) imaging allows real-time visualization of atrial and ventricular anatomy (Fig. 2.10), improving procedural precision while reducing radiation exposure. Randomized and prospective studies consistently show that merging preacquired CT or MR data with EAM enhances operator confidence and can shorten fluoroscopy time without compromising efficacy. For example, early trials demonstrated up to a 35%–40% reduction in fluoroscopy duration and

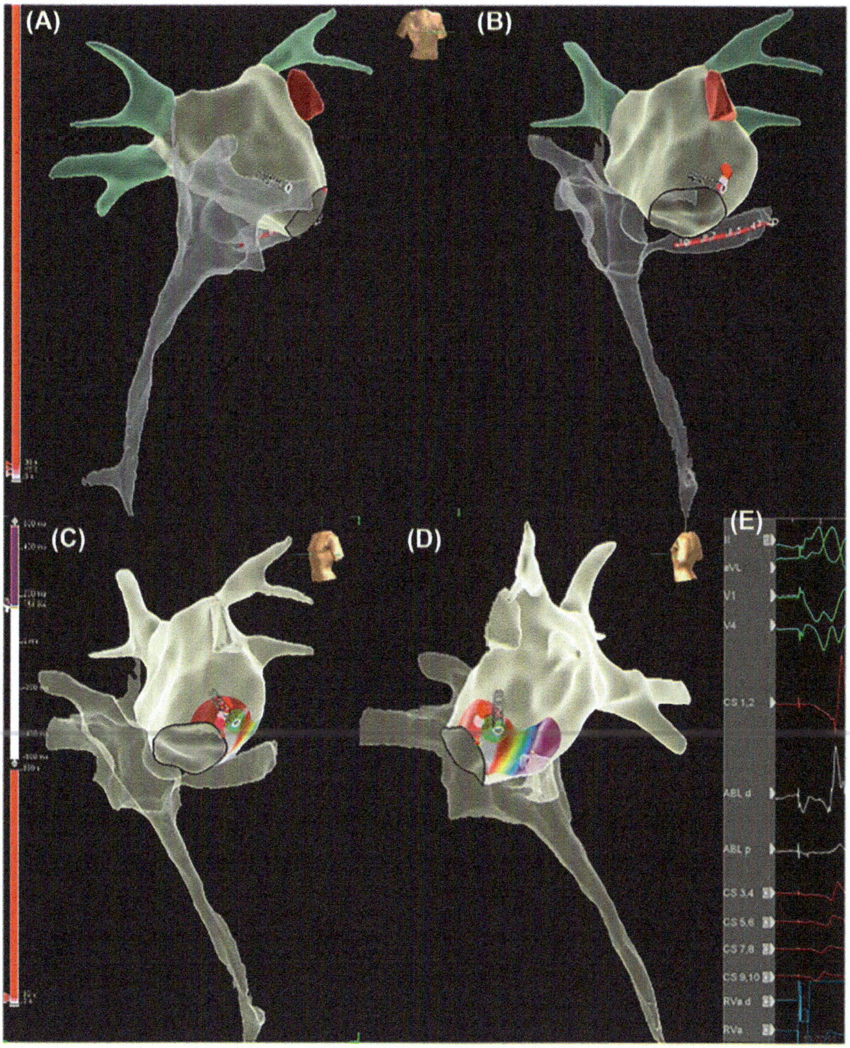

FIGURE 2.10 Current 3D electroanatomic mapping system (EnSite Precision, Abbott Medical Inc.)-guided catheter ablation of a left-sided accessory pathway (fluoro-less).

(A, B) 3D reconstructions of the left and right atria in the right-anterior oblique (A) and left-anterior oblique (B) projections. The ablation catheter is positioned in the area where the accessory pathway potential is recorded, in the lateral mitral annulus region. The rose and red tags indicate the zones where ablation was performed. (C, D) 3D reconstructions of the left and right atria in the left-anterior oblique (C) and lateral (D) views, showing activation mapping during ventricular pacing. The open area indicates the zone of earliest atrial activation through the lateral accessory pathway. The rose and red tags indicate the zones

(*continued*)

maintained 12-month arrhythmia-free survival in atrial fibrillation (AF) ablation compared with standard EAM alone [31,43]. Electrophysiologists are increasingly focused on minimizing radiation exposure for patients, staff, and operators. Through rigorous training and the adoption of advanced imaging and mapping technologies, many centers dedicated to the treatment of complex arrhythmias now perform complex procedures entirely without fluoroscopy. These "zero-fluoro or fluoro-less" approaches have become routine practice in a growing number of highly specialized electrophysiology laboratories worldwide.

The use of 3D electroanatomic mapping guidance (Fig. 2.10) for accessory pathway ablation carries several advantages, including the possibility to identify and annotate the position of the accessory pathway, thus enabling the operator to position the catheter properly to consolidate the ablation lesion after elimination of the accessory pathway and the opportunity to monitor the location of the catheter regarding important anatomic structures that should be avoided (i.e., the atrioventricular node) [44].

Left atrial ablation benefits most clearly from multimodal imaging. High-resolution CT or MR angiography delineates pulmonary vein ostia and complex anatomical variants, guiding lesion sets and minimizing complications such as pulmonary vein stenosis and esophageal injury. Several meta-analyses confirm that image-integrated AF ablation achieves similar or higher single-procedure success (≈75%–85% at 1 year) while lowering radiation exposure by 25%–50% relative to fluoroscopy-only guidance (Fig. 2.11). Furthermore, CMR-derived scar maps fused with EAM enable tailored substrate modification for persistent AF and atrial tachycardia, particularly in redo cases with extensive fibrosis [44,46].

In VT ablation, late-gadolinium-enhanced CMR provides unparalleled characterization of scar heterogeneity and border-zone channels (Fig. 2.12). Prospective series show that preprocedural CMR integration reduces VT recurrence from ~48% to ~32% at one year and decreases procedure duration by enabling targeted substrate ablation [47]. Where CMR is contraindicated, contrast-enhanced multi-detector CT can approximate scar distribution and has demonstrated comparable procedural benefit.

Recent advances in AF ablation imaging include intracardiac echocardiography (ICE)-based 3D reconstruction, real-time rotational angiography (3DRA) with fluoroscopy overlay, and full image integration modules. These methods allow dynamic or continuously updated chamber geometry, enabling minimal- or zero-fluoroscopy workflows. Multiple studies have indicated that such approaches are

where ablation was performed. (E) Intracavitary electrograms during ventricular pacing, with evidence of ventriculoatrial fusion and Kent potential (arrow) in the lateral mitral annulus, as recorded by the ablation catheter.

Reprinted from Bergonti M, et al. Long-term outcomes of near-zero radiation ablation compared with conventional fluoroscopic ablation. Journal of the American College of Cardiology: Electrophysiology. 2021. © 2021 by the American College of Cardiology Foundation. Published by Elsevier. Used with permission (License No. 6226570556301).

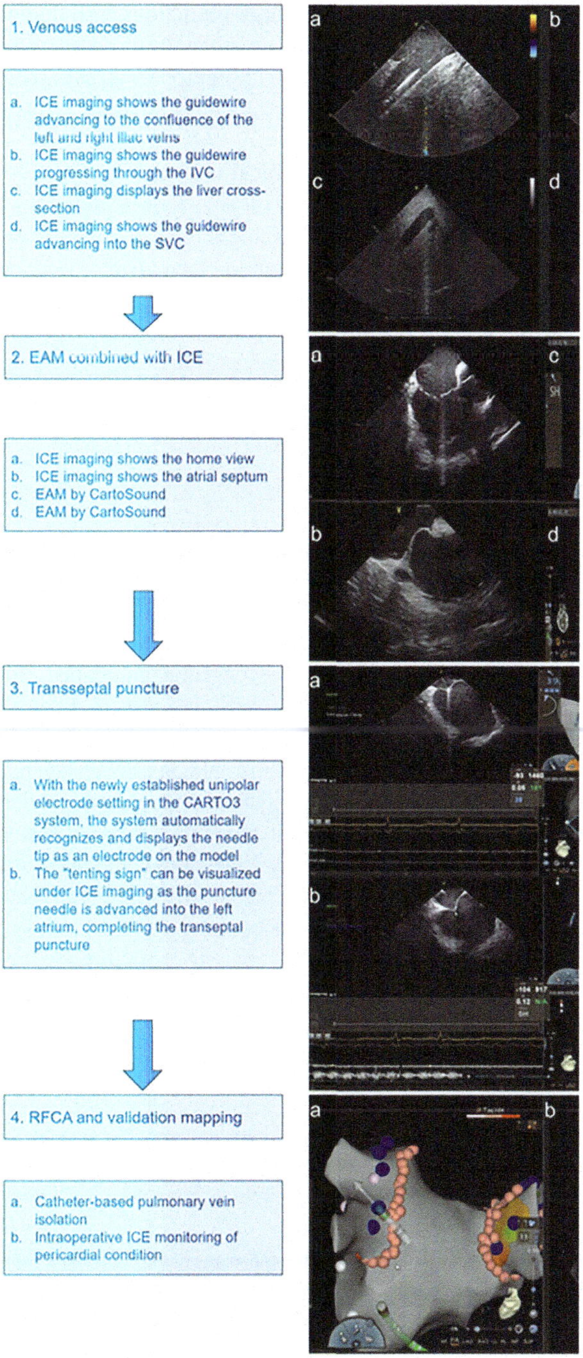
1. Venous access
a. ICE imaging shows the guidewire advancing to the confluence of the left and right iliac veins
b. ICE imaging shows the guidewire progressing through the IVC
c. ICE imaging displays the liver cross-section
d. ICE imaging shows the guidewire advancing into the SVC
2. EAM combined with ICE
a. ICE imaging shows the home view
b. ICE imaging shows the atrial septum
c. EAM by CartoSound
d. EAM by CartoSound
3. Transseptal puncture
a. With the newly established unipolar electrode setting in the CARTO3 system, the system automatically recognizes and displays the needle tip as an electrode on the model
b. The "tenting sign" can be visualized under ICE imaging as the puncture needle is advanced into the left atrium, completing the transeptal puncture
4. RFCA and validation mapping
a. Catheter-based pulmonary vein isolation
b. Intraoperative ICE monitoring of pericardial condition
a
b
c
d
a
c
b
d
a
b
a
b

feasible, safe, and effective in achieving large reductions in radiation exposure for both patients and staff without compromising efficacy [45]. Early feasibility studies of augmented reality (AR) headsets and holographic visualization suggest that immersive 3D representations further enhance spatial orientation without prolonging procedures (Fig. 2.13) [48].

Collectively, these data support the clinical integration of classic and advanced imaging modalities to optimize ablation efficacy, reduce radiation and procedure variability, and expand the electrophysiologist's armamentarium for complex arrhythmias. While large multicenter randomized trials remain limited, the convergence of EAM, high-resolution CT/CMR, and emerging virtual/AR platforms represents the current state-of-the-art in EP imaging and sets the stage for next-generation precision ablation [48].

Key contemporary studies supporting functional and virtual 3D imaging in electrophysiology are summarized in Table 2.4.

2.8 How functional and virtual medical 3D imaging changes practice in electrophysiology [30,31,33,40,42,49]

Preprocedural planning: obtain high-quality CT or CMR when anatomy or scar mapping is expected to change your strategy (congenital anomalies, redo AF, complex left atrial anatomy, suspected epicardial VT substrate). Ensure imaging protocols are optimized for post-processing (thin slices, appropriate contrast timing).

Registration and verification: always perform intraprocedural landmark checks after registration (e.g., coronary ostia, CS ostium, pulmonary vein ostia, valves). Know your system's reported registration error and visually confirm catheter-to-anatomy congruence.

Radiation reduction: adopt fluoroscopy integration modules and EAM-based navigation to reduce fluoroscopy time; this is particularly useful for repetitive procedures and to lower staff and patient dose. Robust local protocols (and training) are required.

When to prioritize imaging-guided substrate ablation: In scar-related VT (especially nonischemic dilated cardiomyopathy or complex postinfarct substrates), image-guided lesion sets (scar homogenization, channel ablation) informed by

FIGURE 2.11

Zero-fluoroscopy workflow. Zero-fluoroscopy workflow using ICE combined with EAM. *ICE*, Intracardiac echocardiography; *EAM*, electroanatomic mapping; *IVC*, inferior vena cava; *SVC*, superior vena cava.

Reprinted from Liu G, Shao B, Wu W, Zhou L, Cui J, Chen W, Zhang R, Liu F. Safety and efficacy of intracardiac echocardiography-guided zero-fluoroscopy ablation in atrial fibrillation patients: a comparative study of high-power short-duration and low-power long-duration strategies. Frontiers in Cardiovascular Medicine. 11:1510889. doi: 10.3389/fcvm.2024.1510889.

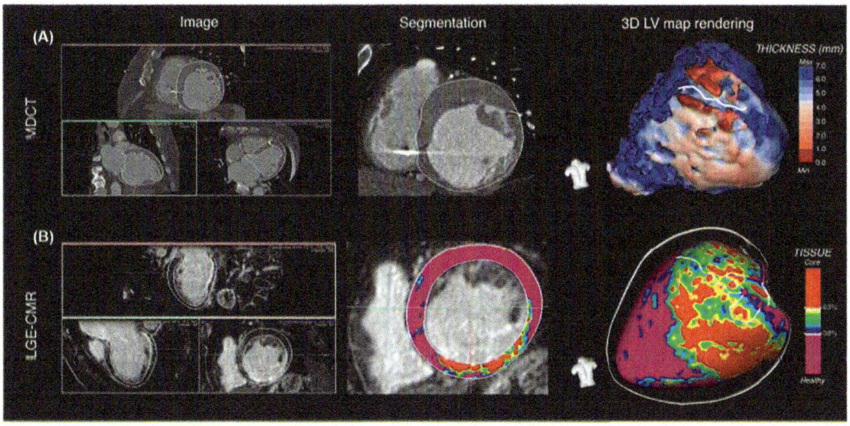

FIGURE 2.12

(A) Preprocedural MDCT-derived image data post-processing and 3D LV wall thinning map rendering with computed tomography channel detection in a patient with ischemic cardiomyopathy who experienced an inferior myocardial infarction. (B) Preprocedural CMR-derived image data post-processing and pixel signal intensity map rendering with detection of BZCs in the same patient. *LGE-CMR*, late gadolinium enhancement cardiac magnetic resonance; *LV*, left ventricle; *MDCT*, multidetector computed tomography.

Reprinted from Penela D, Soto-Iglesias D, Fernández-Armenta J, et al. Outcomes of ventricular tachycardia ablation facilitated by preprocedural late gadolinium enhancement Cardiac Magnetic Resonance (CMR) integration. Europace. 2025;27(4):807–8756. doi:10.1093/europace/euaf051. CC BY 4.0, Copyright © 2025, © The Author(s) 2025. Published by Oxford University Press on behalf of the European Society of Cardiology.

CT/CMR mapping provide higher confidence in targeting and may reduce recurrence; consider in centers with image-fusion expertise.

Use AR/VR cautiously but proactively: trial AR/VR first in planning and teaching; move to intraprocedural adoption only after local simulation and acceptance testing. Regulatory-cleared systems are now available; they involve institutional device/IT (information technology) oversight for data flow and cybersecurity.

2.9 Regulatory, ethical, and economic considerations

Clinicians must understand local regulatory requirements for AI and AR/VR tools [50]. Key ethical issues include patient consent, data privacy, and algorithmic bias. Cost–benefit analyses should consider procedural efficiency and reduced complications. AI's transformative role in interventional cardiology enhances diagnostic accuracy, procedural outcomes, remote monitoring, and education, while acknowledging the need for ethical considerations and a more profound understanding of AI mechanisms in healthcare [50].

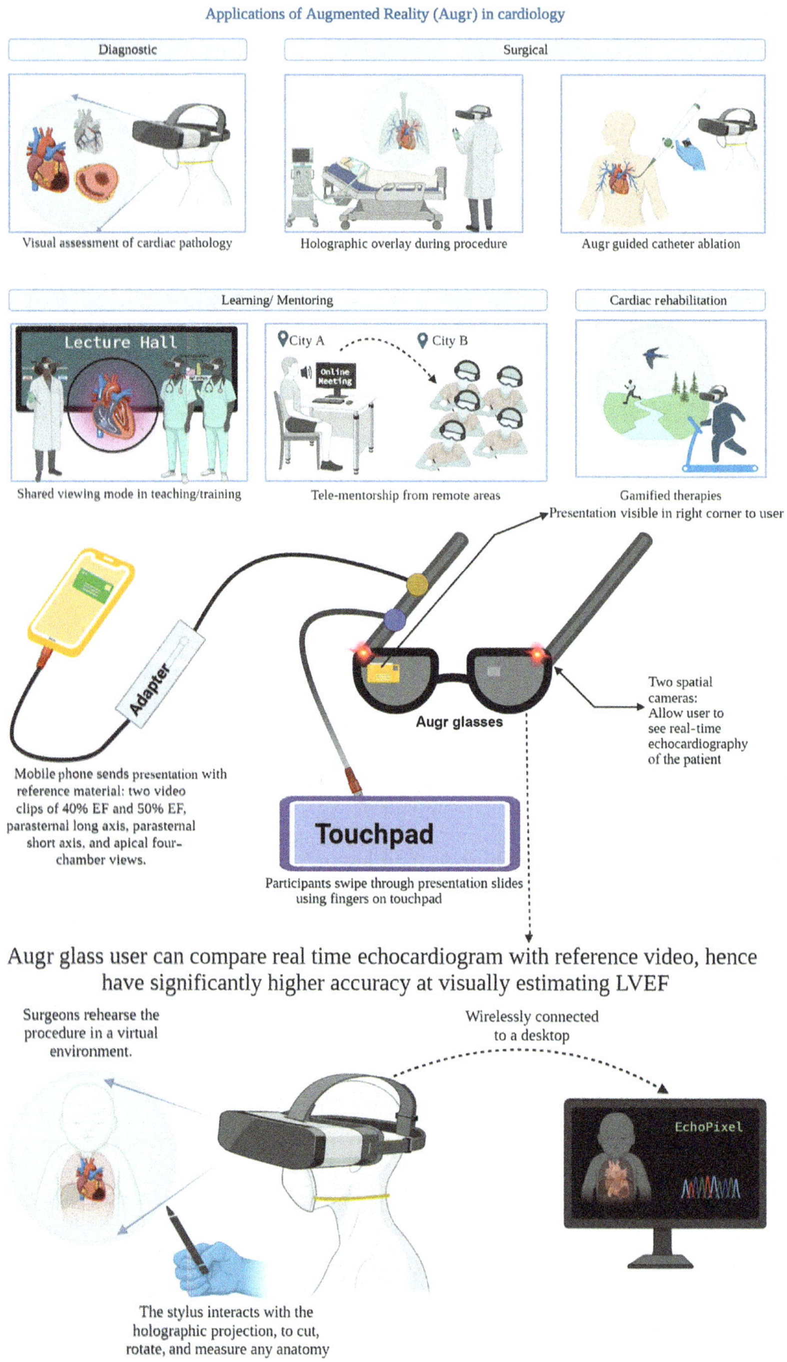
Applications of Augmented Reality (Augr) in cardiology
Diagnostic
Surgical
Visual assessment of cardiac pathology
Holographic overlay during procedure
Augr guided catheter ablation
Learning/ Mentoring
Cardiac rehabilitation
Lecture Hall
City A
Online Meeting
City B
Shared viewing mode in teaching/training
Tele-mentorship from remote areas
Gamified therapies
Presentation visible in right corner to user
Adapter
Augr glasses
Two spatial cameras: Allow user to see real-time echocardiography of the patient
Mobile phone sends presentation with reference material: two video clips of 40% EF and 50% EF, parasternal long axis, parasternal short axis, and apical four-chamber views.
Touchpad
Participants swipe through presentation slides using fingers on touchpad
Augr glass user can compare real time echocardiogram with reference video, hence have significantly higher accuracy at visually estimating LVEF
Surgeons rehearse the procedure in a virtual environment.
Wirelessly connected to a desktop
EchoPixel
The stylus interacts with the holographic projection, to cut, rotate, and measure any anatomy

2.10 What to watch out for

- Underestimation of volumes: 3DE tends to report lower EDV/ESV compared to CMR; important when using thresholds for therapy.
- Image quality variations (patient habitus, arrhythmia, respiratory motion).
- Radiation and contrast exposure in CT; gadolinium risks in MRI (especially in compromised renal function).
- Costs and access: not all centers have CT-FFR or advanced MRI tools.
- Interpretation must always consider clinical context; automated or semiautomated analyses need oversight.

2.11 Limitations and future directions [25,33,35,39,41,48,49]

Despite substantial progress, the integration of advanced 3D imaging into routine care faces challenges: cost, accessibility, and the need for standardized protocols across platforms and vendors. Furthermore, educational gaps persist in training clinicians to interpret and integrate these multidimensional datasets that usually are reserved for selected EP centers. Potential pitfalls include overreliance on automated outputs, poor image quality, and segmentation errors. Future directions involve seamless multimodal fusion, explainable AI, and personalized interventional simulations.

Future directions point toward the broader integration of AI for real-time image acquisition and interpretation, progressive automation of imaging workflows, and the incorporation of virtual and augmented reality platforms for education and preprocedural planning. The ***Jagiellonian University Medical College and its Center for Digital Medicine and Robotics (Kraków, Poland)*** are at the forefront of these innovations, serving as pioneers in a transformative field poised to redefine the way medical education and procedural preparation are conducted (Fig. 2.14) [50,52–54].

FIGURE 2.13 Applications of augmented reality in cardiology.

Virtual/AR platforms represent the current state-of-the-art in EP imaging and set the stage for next-generation precision ablation. The integration of EAM, high-resolution CT/CMR, and emerging virtual/AR platforms allows the physician to have a visual assessment of the cardiac pathology in a more 3D, spatial, and comprehensive way than conventional imaging.

Adapted from Ismail Sharieff I, Smith RL, Wang Y, Patel AM, Gómez J, Edwards MK, Chen Z, Tan R, Almeida J, Brown P, Li H. Applications of augmented reality in cardiology to 2024. Frontiers in Virtual Reality. 2025;2:1580619. https://doi.org/10.3389/frvir.2025.1580619.

Table 2.4 Selected evidence on functional and virtual 3D imaging in electrophysiology.

Study/year	Population and design	Imaging modality/ integration	Primary outcome	Key numeric results	References
Martinek et al., 2007	80 pts with paroxysmal & persistent AF; prospective randomized	CT left atrial angiography integrated with CARTO EAM ("Carto-MERGE")	Freedom from AF at 12 months	85.1% versus 67.9% success (CT-integrated vs. standard EAM, $p<0.05$)	[31]
Kistler et al., 2008	100 pts persistent AF; RCT	CT-EAM integration	Recurrence at 12 months	74% versus 69% (NS), but fewer repeat procedures	[11]

Abbreviations: *AF*, Atrial fibrillation; *EAM*, electroanatomic mapping; *NS*, not significant.

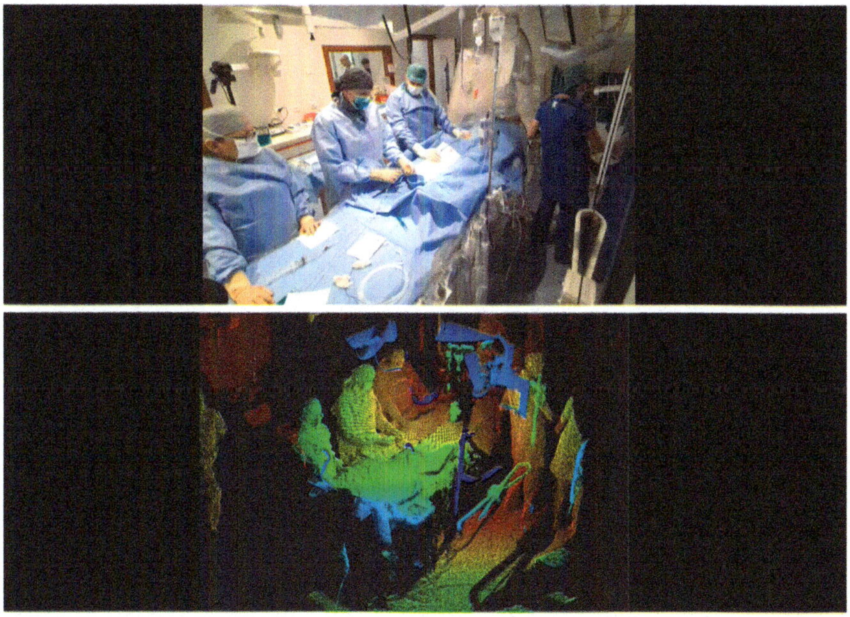

FIGURE 2.14 The use of mixed reality technology during structural surgery.

Professor Klaudia Proniewska-van Dam and her team were working on a 3D holographic room, capturing a 3D operating room in an immersive context. The picture shows an interventional procedure in the middle of complex structural operations. These scenes were created using a spatial awareness system, the Mixed Reality Toolkit (MRTK) based on the HoloMed project [51].

In cardiology:

- Wider availability of fast CT-FFR/AI-based tools to make functional/anatomic testing more accessible.
- Improved intracardiac 3DE hardware and software to reduce underestimation bias, especially in older/disease populations.
- Broader use of 4D flow in MRI, and possibly emerging MRI sequences with faster acquisition, motion compensation [21].
- Use of virtual or augmented reality to view 3D anatomy in clinical or procedural planning.
- Cost-effectiveness studies and outcome trials to further validate that using 3D imaging changes hard outcomes (death, MI, rehospitalization, and recurrences) and justify costs.

In electrophysiology:

- Merging classic electrophysiologic techniques with functional 3D imaging and virtual visualization has moved from experimental to practical in many centers. CT and CMR provide complementary anatomical and substrate

Table 2.5 Conclusions and take-home-messages.

Key point	Take-home message	Supporting references
Substrate targeting	MRI/CT scar integration improves localization of arrhythmogenic substrate, especially in ventricular tachycardia (VT) ablation, supporting scar homogenization and channel ablation strategies.	[41]
Outcome impact	Evidence shows reduced recurrence in select populations (e.g., nonischemic cardiomyopathy, redo procedures), but RCTs show mixed results in routine AF ablation.	[31]
Technology adoption	AR/VR and extended reality are emerging as tools for preprocedural planning, operator training, and eventually intraprocedural guidance.	[48]
Future direction	Integration is best used selectively (complex AF, VT, congenital disease). Next-generation workflows will combine zero-fluoro ICE, CMR scar imaging, and AR/VR visualization.	[45]

information that, when accurately registered into electroanatomic maps, improves procedural planning and may reduce radiation exposure and targeting error.

- In scar-related VT and complex AF, several contemporary series show improved substrate localization and procedural metrics with image fusion; AR/VR platforms are emerging as practical tools for planning and teaching.
- Nevertheless, benefit is operator- and workflow-dependent; rigorous registration checks, institutional quality assurance, and selective use in cases where imaging changes management are essential.

2.12 Recommendations and "When to order" tips

Order CTA + CT-FFR when the patient has stable chest pain, and intermediate pretest likelihood to avoid unnecessary angiography.

Order CMR when myocarditis, infiltrative disease, hypertrophic cardiomyopathy, or where scar burden/tissue characterization will influence treatment.

Use 3DE for valve disease evaluation, especially for mitral/tricuspid, or when 2D echo images are suboptimal.

For monitoring changes over time (e.g., response to therapy), stick with the same 3D imaging modality to avoid measurement bias.

2.13 Recommendations and checklists

- 3D transesophageal echocardiography (TEE) structural heart planning: include full-volume en face views, 3D color Doppler, annular dimensions, and commissural landmarks.
- CT Structural Heart Protocol: ECG-gated thin-slice reconstruction with center-line analysis and metal artifact reduction.
- CMR 3D Evaluation: Cine 3D, contrast LGE 3D volume, T1/T2 mapping, and 4D flow sequences with standardized postprocessing [21].

2.14 Take-home messages (Table 2.5)

Image integration improves safety, efficiency, and radiation reduction.

Benefits are clearest in complex substrates (redo AF, VT, congenital disease).

Long-term outcome data remain mixed, requiring ongoing randomized evidence.

2.15 Conclusions (Table 2.5)

Future practice will likely combine CT/CMR scar imaging, ICE-guided fluoroscopy-free workflows, and AR/VR for enhanced visualization.

- Advanced 3D imaging has become an essential tool in modern cardiology, offering superior accuracy in structural and functional assessment compared to 2D imaging.

- Clinical integration of CT, MRI, 3DE, and hybrid techniques improves diagnostic accuracy, procedural planning, and patient outcomes across cardiology subspecialties.

- In electrophysiology, combining classic methods (fluoroscopy, EAM) with advanced imaging (CT, CMR, ICE, AR/VR) enables safer, more precise, and lower-radiation ablation workflows.

- Limitations include cost, availability, registration errors, and the need for specialized training. Clinical benefits are context- and operator-dependent.

- Future directions point toward AI-driven automation, broader adoption of AR/VR for planning and teaching, and multicenter trials to establish outcome benefits.

Imaging integration has evolved from an experimental adjunct to a practical tool in modern electrophysiology and cardiology. Its greatest strengths lie in improving safety through radiation reduction, enhancing procedural efficiency in complex ablations and complex interventional cases, refining arrhythmia substrate targeting with advanced CT and CMR integration, and improving diagnostic capabilities in difficult cardiac pathologies and congenital cardiac diseases. While outcome improvements are not universal, particularly in first-time AF ablation, clear benefits exist in redo procedures, ventricular tachycardia ablation, and in patient populations with complex anatomy. Current limitations include registration errors, resource requirements, and variable availability across institutions. Future practice will increasingly incorporate multimodality imaging, zero-fluoroscopy workflows, and extended reality platforms. Ultimately, careful integration of these modalities into daily workflows has the potential to improve patient outcomes, optimize procedural strategies, and set new standards in clinical cardiology, evidence-based EP practice, and interventional cardiology procedures.

Acknowledgments

The authors would like to thank Kaja Taton for preparing the drawings for this article. Her contributions improved the clarity and readability of the presented content.

References

[1] Daubert MA, Tailor T, James O, et al. Multimodality cardiac imaging in the 21st century: evolution, advances and future opportunities for innovation. The British Journal of Radiology 2021;94(1117)):20200780. Available from https://doi.org/10.1259/bjr.20200780.

[2] Staub NL, et al. Preoperative three-dimensional strain imaging identifies reduction in left ventricular function and predicts outcomes after cardiac surgery. Journal of the American Society of Echocardiography 2016;29(2):113–21.

[3] Zamorano JL, Pinto FJ, Solano-López J, Bucciarelli-Ducci C. The year in cardiovascular medicine 2020: imaging. European Heart Journal 2021;42(7):740–9. https://doi.org/10.1093/eurheartj/ehaa1035.

[4] Litjens G, Sengupta PP, et al. Artificial intelligence in cardiovascular imaging: JACC state-of-the-art review. Journal of the American College of Cardiology 2019;73(11):1317–35.

[5] William U, Vanegas D, Duque M. Colombian guidelines of cardiovascular electrophysiology. Colombian Journal of Cardiology (RCC) 2011;18(3):201–95. ISSN: 0120-5633.

[6] William U. Cardiovascular electrophysiology in Colombia in cardiovascular electrophysiology guidelines. Colombian Journal of Cardiology (RCC) 2011;18(3):129–30. Available from https://doi.org/10.1016/S0120-5633(11)70176-0.

[7] Puntmann VO, Nagel E, et al. CMR parametric mapping as a tool for myocardial tissue characterization. Journal of Cardiovascular Magnetic Resonance 2019;21:1–28.

[8] Sköldborg V, Madsen PL, Dalsgaard M, Abdulla J. Quantification of mitral valve regurgitation by 2D and 3D echocardiography compared with cardiac magnetic resonance: a systematic review and meta-analysis. The International Journal of Cardiovascular Imaging 2020;36(2):279–89. Available from https://doi.org/10.1007/s10554-019-01713-7. Epub 2019 Oct 29. PMID: 31664679.

[9] Zhao D, Quill GM, Gilbert K, Wang VY, Houle HC, Legget ME, et al. Systematic comparison of left ventricular geometry between 3D-echocardiography and cardiac magnetic resonance imaging. Frontiers in Cardiovascular Medicine 2021;8:728205. Available from https://doi.org/10.3389/fcvm.2021.728205. PMID: 34616783; PMCID: PMC8488135.
[10] Altiok E, Frick M, Meyer CG, Al Ateah G, Napp A, Kirschfink A, et al. Comparison of two- and three-dimensional transthoracic echocardiography to cardiac magnetic resonance imaging for assessment of paravalvular regurgitation after transcatheter aortic valve implantation. The American Journal of Cardiology 2014;113(11):1859–66. Available from https://doi.org/10.1016/j.amjcard.2014.02.038. Epub 2014 Mar 17. PMID: 24837265.
[11] Kistler PM, Rajappan K, Harris S, Earley MJ, Richmond L, Sporton SC, et al. The impact of image integration on catheter ablation of atrial fibrillation using electroanatomic mapping: a prospective randomized study. European Heart Journal 2008;29(24):3029–36. Available from https://doi.org/10.1093/eurheartj/ehn453. Epub 2008 Oct 17. PMID: 18931059.
[12] Huang W, Bai Y, Yang D, et al. Meta-analysis of pulmonary vein isolation ablation for atrial fibrillation: conventional vs low- and zero-fluoroscopy approaches. Journal of Cardiovascular Electrophysiology 2020;31:1754–67. Available from https://doi.org/10.1111/jce.14450.
[13] Debreceni I, Janosi A, Bocz B, et al. Zero fluoroscopy catheter ablation for atrial fibrillation: a systematic review and meta-analysis. Frontiers in Cardiovascular Medicine 2023;10:1178783. Available from https://doi.org/10.3389/fcvm.2023.1178783.
[14] Soto-Iglesias D, Penela D, Jáuregui B, et al. Cardiac magnetic resonance-guided ventricular tachycardia substrate ablation. Journal of American College of Cardiology Electrophysiology 2020;6(4):436–47. Available from https://doi.org/10.1016/j.jacep.2019.11.004.
[15] Biglino G, Capelli C, Wray J, et al. 3D-manufactured models of congenital heart disease for communication in clinical practice: feasibility and acceptability. BMJ Open 2015;5(4):e007165. Available from https://doi.org/10.1136/bmjopen-2014-007165.
[16] Kaplan RM, Narang A, Gay H, Gao X, Gibreal M, Arora R, et al. Use of a novel 4D intracardiac echocardiography catheter to guide interventional electrophysiology procedures. Journal of Cardiovascular Electrophysiology 2021;32(12):3117–24. Available from https://doi.org/10.1111/jce.15251. Epub 2021 Sep 28. PMID: 34554627.
[17] Lang RM, Badano LP, Mor-Avi V, et al. Recommendations for cardiac chamber quantification by echocardiography in adults: an update from the American Society of Echocardiography and the European Association of Cardiovascular Imaging. European Heart Journal of Cardiovascular Imaging 2015;16(3):233–70. Available from https://doi.org/10.1093/ehjci/jev014.
[18] Dissabandara T, Lin K, Forwood M, Sun J, et al. Validating real-time three-dimensional echocardiography against cardiac magnetic resonance, for the determination of ventricular mass, volume, and ejection fraction: a meta-analysis. Clinical Research in Cardiology: Official Journal of the German Cardiac Society 2023;113(4):367–92. Available from https://doi.org/10.1007/s00392-023-02204-5.
[19] Narayan SM, Krummen DE, Shivkumar K, Clopton P, Rappel WJ, Miller JM. Treatment of atrial fibrillation by the ablation of localized sources: CONFIRM (Conventional Ablation for Atrial Fibrillation With or Without Focal Impulse and

Rotor Modulation) trial. Journal of the American College of Cardiology 2012;60(7):628–36. Available from https://doi.org/10.1016/j.jacc.2012.05.022. Epub 2012 Jul 18. PMID: 22818076; PMCID: PMC3416917.

[20] Garg P, Swift AJ, Zhong L, Carlhäll CJ, Ebbers T, Westenberg JJM, et al. Assessment of intracardiac flow and hemodynamics by 4D flow cardiovascular magnetic resonance: a review of current applications, challenges and future perspectives. Journal of Cardiovascular Magnetic Resonance: Official Journal of the Society for Cardiovascular Magnetic Resonance 2020;22:87.

[21] Dyverfeldt P, Bissell M, Barker AJ, Bolger AF, Carlhäll CJ, Ebbers T, et al. 4D Flow cardiovascular magnetic resonance consensus statement: 2023 update. Journal of Cardiovascular Magnetic Resonance 2023;25:78. Available from https://doi.org/10.1186/s12968-023-00957-8.

[22] Disertori M, Rigoni M, Pace N, Casolo G, Masè M, Gonzini L, et al. Myocardial fibrosis assessment by LGE imaging in arrhythmic and non-arrhythmic patients: a meta-analysis. Journal of the American College of Cardiology 2019;74(17):2230–44.

[23] Dorbala S, Ananthasubramaniam K, Armstrong IS, et al. Single photon emission computed tomography (SPECT) myocardial perfusion imaging guidelines: instrumentation, acquisition, processing, and interpretation. Journal of Nuclear Cardiology: Official Publication of the American Society of Nuclear Cardiology 2018;25(5):1784–846. Available from https://doi.org/10.1007/s12350-018-1283-y.

[24] Feuchtner GM, Lacaita PG, Bax JJ, Rodriguez F, et al. AI-Quantitative CT coronary plaque features associate with a higher relative risk in women: CONFIRM2 registry. Circulation: Cardiovascular Imaging 2025. Available from https://doi.org/10.1161/CIRCIMAGING.125.018235. **PubMed:** PMID 40162910.

[25] Teske A, Driessen RS, Nørgaard BL, Kwiecinski J, Neefjes LA, et al. Machine learning from quantitative coronary computed tomography angiography predicts fractional flow reserve–defined ischemia and impaired myocardial blood flow. Circulation of Cardiovascular Imaging 2022;15(10):e014369. Available from https://doi.org/10.1161/CIRCIMAGING.122.014369.

[26] Wang J, Wang S, Liang W, Zhang N. Yan Z. The auto-segmentation for cardiac structures using a dual-input deep learning network based on vision saliency and transformer. Journal of Applied Clinical Medical Physics / American College of Medical Physics 2022;23(5):e13597. Available from https://doi.org/10.1002/acm2.13597.

[27] Ballangrud R, Vogel K, Chilloyan D, et al. Diagnostic utility of artificial intelligence for left ventricular scar identification using cardiac magnetic resonance imaging - a systematic review. Heart Rhythm: The Official Journal of the Heart Rhythm Society 2022;19(8):1340–50. Available from https://doi.org/10.1016/j.hrthm.2022.03.011.

[28] Fahmy AS, Rowin EJ, Arafati A, Al-Otaibi T, Maron MS, Nezafat R, et al. Radiomics and deep learning for myocardial scar screening in hypertrophic cardiomyopathy. Journal of Cardiovascular Magnetic Resonance: Official Journal of the Society for Cardiovascular Magnetic Resonance 2022;24:40. Available from https://doi.org/10.1186/s12968-022-00869-x.

[29] Chnafa C, Mendez S, Nicoud F. Application of patient-specific computational fluid dynamics in coronary and intra-cardiac flow simulations: challenges and opportunities. Frontiers in Cardiovascular Medicine 2018;5:66. Available from https://doi.org/10.3389/fcvm.2018.00066.

[30] Özcan EE, Turan ÖE, Yilancioğlu RY, Inevi U, Akdemir B. Combining conventional technique with fluoroscopy integration module in accessory pathway ablation. Journal of Cardiovascular Electrophysiology 2023;34(12):2573–80. Available from https://doi.org/10.1111/jce.16110. PMID: 37890039.
[31] Martinek M, Nesser HJ, Aichinger J, Boehm G, Purerfellner H. Impact of integration of multislice computed tomography imaging into 3-dimensional electroanatomic mapping on clinical outcomes, safety, and efficacy using radiofrequency ablation for atrial fibrillation. Pacing and Clinical Electrophysiology: PACE 2007;30(10):1215–23. Available from https://doi.org/10.1111/j.1540-8159.2007.00843.x. PMID: 17897124.
[32] Kistler PM, Earley MJ, Harris S, Abrams D, Ellis S, Sporton SC, et al. Validation of three-dimensional cardiac image integration: use of integrated CT image into electroanatomic mapping system to perform catheter ablation of atrial fibrillation. Journal of Cardiovascular Electrophysiology 2006;17(4):341–8. Available from https://doi.org/10.1111/j.1540-8167.2006.00371.x. PMID: 16643352.
[33] Patel AM, Heist EK, Chevalier J, Holmvang G, D'Avila A, Mela T, et al. Effect of presenting rhythm on image integration to direct catheter ablation of atrial fibrillation. Journal of Interventional Cardiac Electrophysiology: An International Journal of Arrhythmias and Pacing 2008;22(3):205–10. Available from https://doi.org/10.1007/s10840-008-9265-7. Epub 2008 May 28. PMID: 18506606.
[34] Liao SC, Shao SC, Gao SY, et al. Augmented reality visualization for ultrasound-guided interventions: a pilot randomized crossover trial to assess trainee performance and cognitive load. BMC Medical Education 2024;24:1058. Available from https://doi.org/10.1186/s12909-024-05998-8.
[35] Liu SX, Zhang Y, Zhang XW. Impact of image integration on catheter ablation for atrial fibrillation using three-dimensional electroanatomic mapping: a meta-analysis. Pacing and Clinical Electrophysiology: PACE 2012;35(10):1242–7. Available from https://doi.org/10.1111/j.1540-8159.2012.03492.x. Epub 2012 Aug 20. PMID: 22906107.
[36] Bertaglia E, Bella PD, Tondo C, Proclemer A, Bottoni N, De Ponti R, et al. Image integration increases efficacy of paroxysmal atrial fibrillation catheter ablation: results from the CartoMerge Italian Registry. Europace: European Pacing, Arrhythmias, and Cardiac Electrophysiology: Journal of the Working Groups on Cardiac Pacing, Arrhythmias, and Cardiac Cellular Electrophysiology of the European Society of Cardiology 2009;11(8):1004–10. Available from https://doi.org/10.1093/europace/eup131.
[37] Ouyang F, Tilz R, Chun J, Schmidt B, Wissner E, Zerm T, et al. Long-term results of catheter ablation in paroxysmal atrial fibrillation: lessons from a 5-year follow-up. Circulation 2010;122(23):2368–77. Available from https://doi.org/10.1161/CIRCULATIONAHA.110.946806.
[38] Steven D, Rostock T, Lutomsky B, Klemm H, Servatius H, Drewitz I, et al. Reduced fluoroscopy during atrial fibrillation ablation: benefits from remote magnetic navigation. Europace: European Pacing, Arrhythmias, and Cardiac Electrophysiology: Journal of the Working Groups on Cardiac Pacing, Arrhythmias, and Cardiac Cellular Electrophysiology of the European Society of Cardiology 2008;10(5):583–7. Available from https://doi.org/10.1093/europace/eun048.
[39] Roca-Luque I, Mont-Girbau L. Cardiac magnetic resonance for ventricular tachycardia ablation and risk stratification. Frontiers in Cardiovascular Medicine 2022;8:797864. Available from https://doi.org/10.3389/fcvm.2021.797864.

[40] Deneke T, Kutyifa V, Hindricks G, Sommer P, Zeppenfeld K, Carbucicchio C, et al. Pre-and post-procedural cardiac imaging (computed tomography and magnetic resonance imaging) in electrophysiology: a clinical consensus statement of the European Heart Rhythm Association and European Association of Cardiovascular Imaging of the European Society of Cardiology. Europace: European Pacing, Arrhythmias, and Cardiac Electrophysiology: Journal of the Working Groups on Cardiac Pacing, Arrhythmias, and Cardiac Cellular Electrophysiology of the European Society of Cardiology 2024;26(5):euae108. Available from https://doi.org/10.1093/europace/euae108. PMID: 38743765; PMCID: PMC11104536.

[41] Andreu D, Berruezo A, Ortiz-Perez JT, Silva E, Mont L, Boussy T, et al. Integration of 3D electroanatomic maps and magnetic resonance scar characterization into the navigation system to guide ventricular tachycardia ablation. Circulation: Arrhythmia and Electrophysiology 2011;4(5):674–83. Available from https://doi.org/10.1161/CIRCEP.111.963124.

[42] Abbara S, Blanke P, Maroules CD, Cheezum M, Choi AD, Han BK, et al. SCCT guidelines for the performance and acquisition of coronary computed tomographic angiography: a report of the Society of Cardiovascular Computed Tomography Guidelines Committee. Journal of Cardiovascular Computed Tomography 2016;10(6):435–49. Available from https://doi.org/10.1016/j.jcct.2016.10.002.

[43] Balk EM, Garlitski AC, Alsheikh-Ali AA, Terasawa T, Chung M, Ip S. Predictors of atrial fibrillation recurrence after radiofrequency catheter ablation: a systematic review. Journal of Cardiovascular Electrophysiology 2010;21(11):1208–16. Available from https://doi.org/10.1111/j.1540-8167.2010.01798.

[44] Bergonti M, et al. Long-term outcomes of near-zero radiation ablation compared with conventional fluoroscopic ablation. Journal of the American College of Cardiology Electrophysiology 2021.

[45] Liu G, Shao B, Wu W, Zhou L, Cui J, Chen W, et al. Safety and efficacy of intracardiac echocardiography-guided zero-fluoroscopy ablation in atrial fibrillation patients: a comparative study of high-power short-duration and low-power long-duration strategies. Frontiers in Cardiovascular Medicine 2024;11:1510889. Available from https://doi.org/10.3389/fcvm.2024.1510889.

[46] Mammadli A, Demirtola AI, Diker E, et al. Impact of image integration on clinical and procedural outcomes of radiofrequency catheter ablation of atrial fibrillation: a meta-analysis of randomized controlled trials. Journal of Arrhythmia 2021;37(3):550–5.

[47] Penela D, Soto-Iglesias D, Fernández-Armenta J, et al. Outcomes of ventricular tachycardia ablation facilitated by pre-procedural late gadolinium enhancement Cardiac Magnetic Resonance (CMR) integration. Europace: European Pacing, Arrhythmias, and Cardiac Electrophysiology: Journal of the Working Groups on Cardiac Pacing, Arrhythmias, and Cardiac Cellular Electrophysiology of the European Society of Cardiology 2025;27(4):807–8756. Available from https://doi.org/10.1093/europace/euaf051.

[48] Ismail, Sharieff I, Smith RL, Wang Y, Patel AM, Gómez J, et al. Applications of augmented reality in cardiology to 2024. Frontiers in Virtual Reality 2025;2:1580619. Available from https://doi.org/10.3389/frvir.2025.1580619.

[49] Besson FL, Treglia G, Bucerius J, Anagnostopoulos C, Buechel RR, Dweck MR, et al. A systematic review for the evidence of recommendations and guidelines in hybrid nuclear cardiovascular imaging. European Journal of Nuclear Medicine and Molecular

Imaging 2024;51(8):2247–59. Available from https://doi.org/10.1007/s00259-024-06597-x. Epub 2024 Jan 15. PMID: 38221570; PMCID: PMC11178580.

[50] Rudnicka Z, Pręgowska A, Glądys K, Perkins M, Proniewska K. Advancements in artificial intelligence-driven techniques for interventional cardiology. Cardiology Journal 2024;31(2):321–41. Available from https://doi.org/10.5603/cj.98650.

[51] HoloMed (n.d.) (Holographic MedAssistant 3D operating room with unlimited perspective change and remote support) https://holomed.cm-uj.krakow.pl/.

[52] Proniewska K, Khokhar AA, Dudek D. Advanced imaging in interventional cardiology: mixed reality to optimize preprocedural planning and intraprocedural monitoring. Kardiologia Polska 2021;79(3):331–5. Available from https://doi.org/10.33963/KP.15814.

[53] Pociask E, Proniewska K, Bruining N. The potential role of intracoronary imaging at the dawn of the fourth revolution in interventional cardiology. Revista Argentina de Cardioangiología Intervencionista 2016;7(1):15–22. Available from https://doi.org/10.30567/RACI/201601/0015-0022.

[54] Proniewska K, Pręgowska A, Walecki P, Dołęga-Dołęgowski D, Ferrari R, Dudek D. Overview of the holographic-guided cardiovascular interventions and training-a perspective. Bio-Algorithms and Med-Systems 2020;16(3). Available from https://doi.org/10.1515/bams-2020-0043. 20200043-1-9.

CHAPTER

Coronary arteries computed tomography

3

Łukasz Brandt[1,2] and Paweł Banyś[1]

[1]*Functional and Virtual Medical 3D Imaging Laboratory, Department of Diagnostic Imaging, University Hospital in Kraków, Kraków, Poland,*

[2]*Department of Electroradiology, Jagiellonian University Medical College, Kraków, Poland*

The evolution from early computed tomography (CT) systems, which required several minutes to acquire limited brain images, to modern multislice CT (MSCT) scanners has dramatically improved imaging speed and quality. Innovations such as slip-ring technology, fan-shaped X-ray beams, and advanced detector arrays have enabled rapid spiral scanning. Currently, MSCT is a reliable, noninvasive diagnostic tool, particularly valuable for assessing coronary arteries. Techniques such as breath-holding and electrocardiogram (ECG)-gated acquisition reduce respiratory and cardiac motion, while temporal resolution depends largely on gantry rotation speed and partial data reconstruction.

Despite these improvements, challenges such as motion artifacts, calcifications, and metallic stents persist in coronary CT angiography. Dual-source CT (DSCT), incorporating two X-ray tubes and detectors positioned at a 90-degree offset, addresses these limitations by significantly enhancing temporal resolution and reducing acquisition time.

Further progress is represented by photon-counting CT (PCCT), which introduces direct conversion detectors characterized by low electronic noise and energy-resolved signal detection. This technology enables ultra-high-spatial resolution (down to 0.2 mm) and reduced radiation dose, while achieving temporal resolutions as low as 66 ms in cardiac imaging.

Cardiac CT examinations include noncontrast studies for calcium scoring and contrast-enhanced angiography for comprehensive assessment of the heart's structure and vasculature. Prospective and retrospective ECG gating techniques are selected based on heart rate and rhythm to optimize image quality and radiation exposure.

3.1 Introduction

The rapid technological and information technology developments observed in recent years in the design of CT scanners, image data acquisition, and processing

Functional and Virtual Medical 3D Imaging in Cardiovascular Medicine. DOI: https://doi.org/10.1016/B978-0-443-44570-5.00009-4

allow the acquisition of CT images with high-temporal and spatial resolution synchronized with the ECG signal. Therefore, scanners, after patient preparation to the recommended heart rate, allow imaging of a moving organ, such as the heart, within several seconds, thereby avoiding artifacts. CT is one of the most powerful nondestructive techniques and provides morphological and physical information about the internal structure of the objects under investigation. Based on the acquisition of digital X-ray images, enabling the reconstruction of a 3D volume and the generation of two-dimensional tomographic slices. Since its inception, CT imaging has represented a breakthrough in medicine. Over the past several years, the role of CT in cardiovascular diagnosis has become increasingly prominent [1].

Initial CT scanners were limited to topographic sections of the brain, and images took several minutes to acquire. Slip-ring technology replaced the cables of older scanners and permitted fast spiral scanning. The scan speed was further improved by changing the shape of the radiation beam (to a fan shape) and increasing the number and quality of X-ray detectors in what are called multislice CT MSCT scanners.

Now, MSCT is a reliable noninvasive diagnostic method, especially useful for examinations of the heart and coronary arteries. Respiratory motion is excluded by performing the scan during a breath-hold, and cardiac motion is minimized by acquiring images when the heart is least mobile during diastole. The temporal resolution of CT scanners is essentially determined by the gantry rotation speed around the patient. It is possible to accurately reconstruct images using data acquired from a 180-degree rotation. CT angiography of the coronary arteries permits assessment; however, image quality is still impaired by motion artifacts, calcifications, and metal artifacts from stents. The newly implemented technique, DSCT, is a promising method for assessing coronary arteries due to its significant improvement in temporal resolution. DSCT is equipped with two X-ray tubes. Two corresponding detectors are oriented in the gantry with an angular offset of 90 degrees. Generally, DSCT technology comprises two X-ray sources and two detectors, used simultaneously in different scanning modes. The two X-ray source/detector systems rotate simultaneously, capturing image data in half the time required with conventional technology.

Advancements in CT design enabled the construction of a PCCT scanner. Thanks to its modern design, the PCCT device's detector enables the direct conversion of X-ray radiation into an electrical signal. The greatest advantage of this type of detector design is its low electronic noise level and the proportional conversion of signal intensity with radiation energy. The CT scanner allows examinations with a minimum slice thickness of 0.2 mm and a reduced radiation dose compared to "traditional" scanners. By using two tube-detector systems, it is possible to achieve a temporal resolution of 66 ms in cardiac examinations.

A CT scan of the heart is performed without a contrast agent, primarily to assess coronary artery calcium (CAC) and aortic valve calcium (AVC). Examination with iodinated contrast coronary computed tomography angiography (CCTA) for heart assessment, including the size of its individual chambers, structural abnormalities, myocardial homogeneity, pulmonary vein, and coronary vessels [2–5].

Two approaches to cardiac gating are typically used: prospective ECG triggering and retrospective ECG gating.

- prospective—sequential scanning prespecified phase of the cardiac cycle, typically 75% after R breakdown:
- stable sinus rhythm and a heart rate of fewer than 60 beats per minute
- prospective ECG triggering high-pitch dual-source protocol
- stable sinus rhythm and a heart rate of fewer than 70 beats per minute
- prospective ECG triggering (step and shoot) protocol
- retrospective—spiral scanning in the full cardiac cycle with dose modulation in the systolic phase
- atrial fibrillation/arrhythmia
- retrospective ECG-gated spiral protocol in the full cardiac cycle

Technology developed with scanners with 64, 128, 256, 320 slices and DSCT. For an MSCT scanner to be capable of cardiac imaging, it must fulfill the following requirements:

- synchronizing with ECG—heart rate <65 bpm
- optimal spatial resolution (slice thickness)—max. 0.6 mm (calcium scoring—3 mm)
- optimal temporal resolution 165 ms to minimize motion artifacts. The gantry rotation speed should be at least 330 ms or even faster.

3.2 Radiation exposure

The radiation dose in cardiac CT depends on protocol selection (retrospective, prospective), scan acquisition parameters (kV, mA, mAs), dose reduction algorithms (tube current modulation), pitch—moving the patient table for one full rotation of the tube-detector system, examination indications, and range of study [6].

3.3 Cardiac computed tomography without contrast

3.3.1 Coronary artery calcium scoring

CAC measured in CT is an established method for assessing the risk of coronary artery disease (Fig. 3.1). X-ray technology was first utilized in cardiology in the early 1990s by Arthur Agatston to evaluate calcifications in coronary arteries with electron-beam CT. The CAC score is obtained using CT with 3 mm axial slices covering the heart, employing ECG gating with a sequential prospective technique (Fig. 3.2). No heart medication to lower the heart rate is necessary. For heart rates exceeding 70 bpm, retrospective gating is recommended. Calcification is defined as at least 1 mm of hyperattenuation in the coronary artery on 3 pixels or more, with a

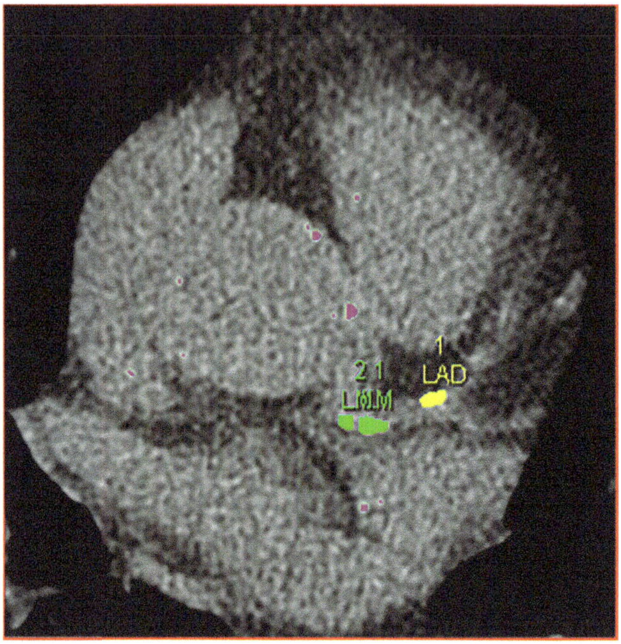

FIGURE 3.1

Coronary artery calcium scoring. Calcification in left main (green color) and left anterior descending (yellow color).

minimum value of 130 Hounsfield units (HU). The CAC value is quantified using the Agatston score, which multiplies the area and volume of calcification by its density. The calcium scoring factor depends on the measured calcium value in HU: 1 for 130 to 199 HU, 2 for 200 to 299 HU, 3 for 300 to 399 HU, and 4 for 400 HU and above. The total calcium score is the sum of the scores for individual arteries (left main—LM, left anterior descending—LAD, circumflex—Cx, and right coronary artery—RCA). The negative predictive value (NPV) of coronary artery calcium scoring is very high at 95% [7,8].

According to the 2007 American College of Cardiology Foundation/American Heart Association (ACC/AHA) guidelines, data from 27,622 asymptomatic patients were analyzed. The relative risk of major adverse cardiovascular events (MACE) was calculated, which included all-cause mortality, cardiac mortality, and nonfatal myocardial infarction. Results are presented in Table 3.1. [9].

Calcium scoring is recommended in high-risk asymptomatic subjects, including:

- diabetic patients in whom coronary CAC has been shown to contribute to higher mortality in comparison to nondiabetic subjects with calcifications and diabetic subjects with calcifications,

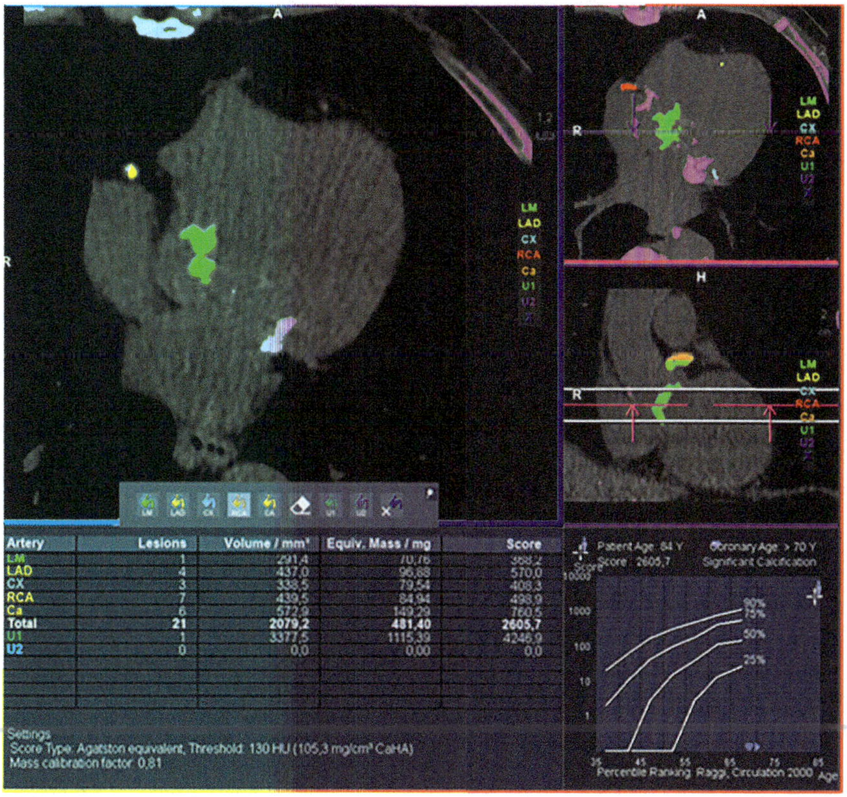

FIGURE 3.2

Aortic valve calcium scoring in green color.

- smokers in whom CAC is significantly more frequent, and the presence of other risk factors increases cardiovascular risk [10].

3.3.2 Aortic valve calcium scoring

In assessing aortic stenosis, when echocardiography is uncertain about severity or dobutamine stress echocardiography is inconclusive or contraindicated, noncontrast cardiac CT calcium scoring can help confirm the severity of aortic stenosis (AS), as it offers a load-independent anatomical assessment. In men, calcium scores greater than 2000 Agatston units and greater than 1200 in women generally indicate severe aortic stenosis (AS) [11].

Table 3.1 Calcium score, interpretation, and future risk of major adverse cardiac events (MACE).

Future risk of major adverse cardiac events (MACE)	
Calcium score	Interpretation/prognosis
0	No calcium, low risk of future MACE
1–99	Average risk; relative risk 1.9% (95% confidence interval 1.3%–2.8%) of future MACE
100–400	Moderate risk; relative risk 4.3% (95% confidence interval 3.1%–6.1%) of future MACE
400–9999	High-risk; relative risk 7.2% (95% confidence interval 4.2%–9.9%) of future MACE
>1000	Very high risk; relative risk 10.8% (95% confidence interval 4.2%–27.7%) of future MACE

3.4 Cardiac computed tomography with contrast

3.4.1 Contrast injection

However, to obtain high- quality images of the heart and coronary vessels, in addition to the general principles of contrast agent use in CT examinations, the planning and execution of the exam must also consider the appropriate selection of the contrast agent, the correct high flow rate (which involves choosing the right cannula) and site of administration, as well as a delay in scanning relative to contrast injection that is appropriate for the specific pathology. The use of dual-tube automatic syringes (one tube filled with contrast media, the other with saline) for contrast delivery has nearly halved the amount of contrast used compared with the single-head systems originally employed for CT scans (Fig. 3.3). Adequate contrast in the examined structures, determined by the iodine concentration in the blood, along with appropriate timing of the scan based on the suspected pathology, ensures the exam's high diagnostic value (Figs. 3.4 and 3.5).Before performing CCTA, an appropriate venous line (most often in the antecubital vein, 18 or 20 gauge) must be established to allow rapid biphasic contrast media injection (0. 6–1 mL per kg of body weight) at a protocol speed of 5–7 mL/sec, followed by 40 mL/sec of saline at the same rate during the scan. Patients are often given nitroglycerin to induce coronary vasodilation and beta-blockers (either orally or intravenously) to achieve a heart rate below 65 bpm, thus optimizing image quality. Contraindications to nitroglycerin include severe aortic stenosis and glaucoma. The scan is performed during a single breath-hold lasting 3 to 12 seconds; the timing for optimal contrast can be determined using bolus tracking, which monitors contrast attenuation in the ascending or descending aorta in real time, or with a "test

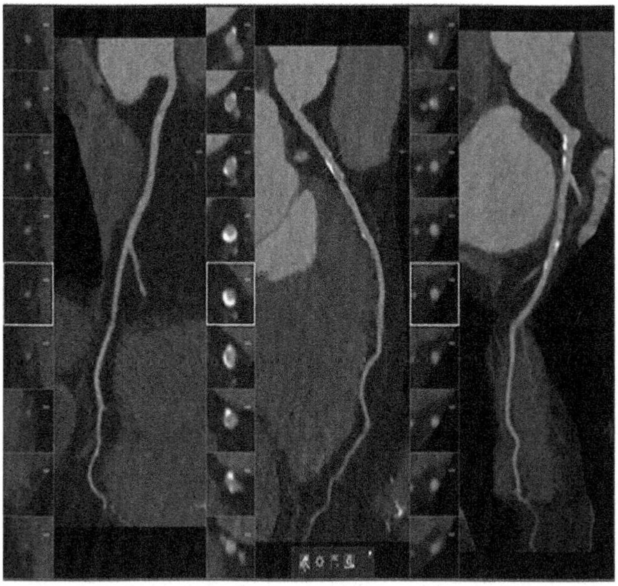

FIGURE 3.3

Curved multiplanar reformation of LAD, circumflex, and left anterior descending.

bolus" of a small amount of contrast (10 mL). Scans are typically performed in the craniocaudal direction (Fig. 3.6); however, when larger areas need imaging or when coronary artery bypass grafts are present, it is possible to scan in the opposite direction relative to the native vessels. Different reconstruction algorithms or "kernels" are used to convert raw scan data into DICOM images. Sharper but noisier kernels are employed to reduce blooming artifacts in calcification and stent visualization [3,4,12].

3.4.2 Preparation and contraindication to scan

- take medicine that drops the heart rate,
- do not use sildenafil (Viagra or Revatio) or any similar medication on the day before or the day of the exam,
- atrial fibrillation,
- severe kidney impairment (eGFR < 30 mL/min/m^2),
- uncontrolled tachycardia,
- obesity (BMI > 40); inability to comply with scan requirements (patients unable to follow instructions).

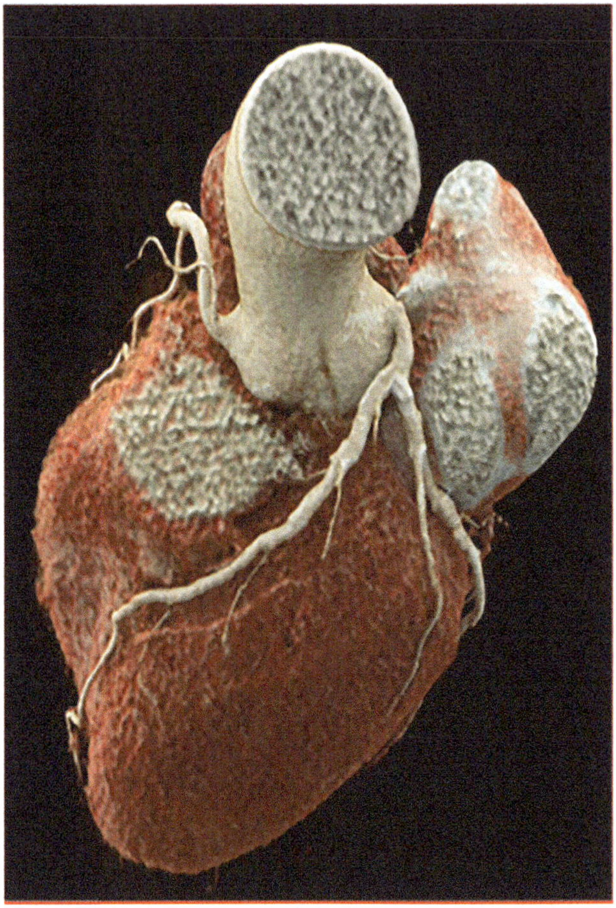

FIGURE 3.4

Volume render reconstruction. Left main, left anterior descending, circumflex, and circumflex.

3.4.3 Indication to cavitary/coronary angiography

- coronary artery examination—exclusion/confirmation of coronary artery disease and anomalies,
- assessment of stent patency (with diameter > 3 mm), procedure planning,
- assessment of bypass patency, procedure planning,
- assessment of pulmonary venous drainage/left atrial appendage before ablation,

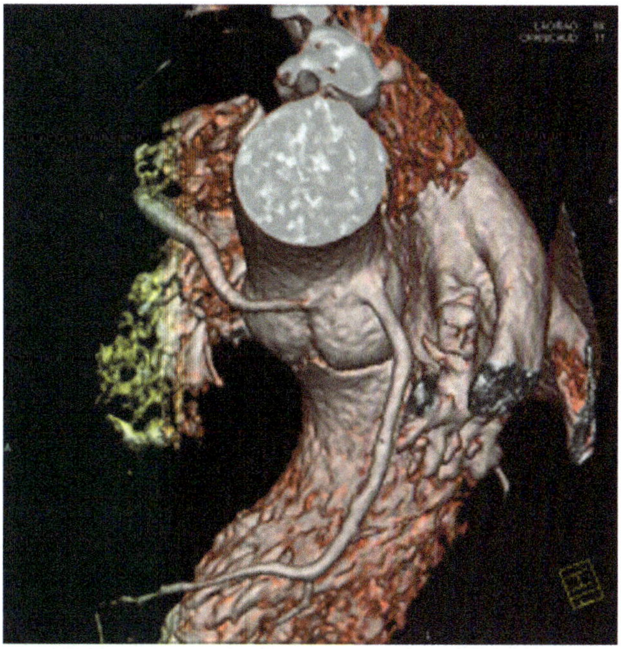

FIGURE 3.5

Volume render reconstruction. Anomaly of origin right coronary artery.

- assessment of the coronary sinus and coronary venous system for resynchronization procedures,
- for the diagnosis of chest pain (triple rule-out protocol),
- cardiac examination of the heart chamber function and dimensions based on retrospective ECG spiral scanning,
- assessment of myocardial morphology,
- assessment of valve morphology (pre and postoperative),
- assessment of cardiac and pericardial tumors,
- assessment of the pericardium, for example, constrictive pericarditis,
- assessment of lead position,
- assessment of the great thoracic vessels,
- assessment before transcatheter aortic valve implantation or other minimally invasive valve procedures (Fig. 3.7),
- assessment of the aorta, including aneurysms and dissection,
- assessment of chronic allograft vasculopathy,
- scar assessment in patients who have contraindications to cardiac magnetic resonance imaging (MRI),
- to plan for open or minimally invasive/robotic heart surgery.

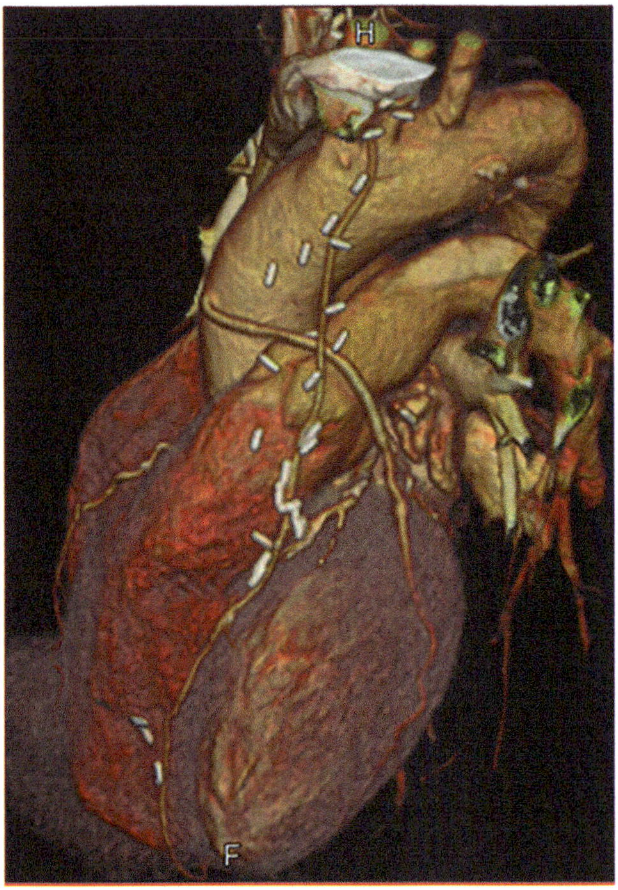

FIGURE 3.6

Coronary artery bypass graft. Patent left internal mammary artery (LIMA) to left anterior descending and venous graft to diagonal (Dg) branch.

Coronary artery CT imaging with contrast enables:

- coronary artery anatomy (anomalies),
- plaque presence,
- plaque location, degree of artery stenosis,
- plaque morphology (length, calcification),
- obstruction (length).

Coronary artery bypass graft CT imaging with contrast enables:

- assessment of arterial and venous graft type:

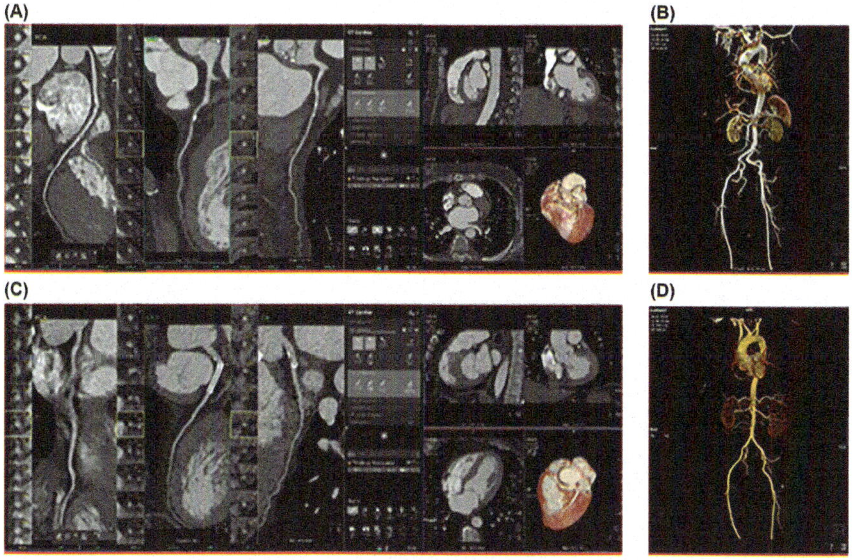

FIGURE 3.7

Computed tomography for transcatheter aortic valve implantation planning. (A) A computed tomography scan of the human body focusing on the aorta and surrounding structures; (B) a 3D reconstruction of the human aorta and renal arteries; (C) similar to A, with cross-sectional views of the aorta, heart, and kidneys; (D) a 3D reconstruction of the human aorta and renal arteries, similar to image B, highlighting the branching of the aorta and the connection to the renal arteries.

- patency,
- stenosis,
- anastomoses,
- assessment of native arteries
- features that facilitate imaging of grafts:
- relatively large graft diameter (especially venous grafts),
- relatively low mobility during the examination,
- features that complicate imaging of bypass grafts:
- presence of metal elements—artifact (metal sutures, clips, markers),
- imaging of distal anastomoses increased calcification of native arteries,
- arrhythmia—phase shifts.

Stent CT imaging with contrast enables:

- assessment of stent patency,
- stent-to-main vessel branch ratio,

- assessment of poststent contrast,
- assessment of stent remodeling (bend, lack of expansion),
- assessment of stent adhesion to the vessel wall,
- assessment of the connection between two adjacent stents,
- complicate imaging of stent:
- presence of artifacts related to stent design
- diameter less than 3 mm,
- presence of calcifications adjacent to the stents [13].

3.4.4 Postprocessing

Image analysis is performed on advanced workstations equipped with dedicated software including image reconstructions:

- transaxial 2-dimensional image stacks multiplanar reformation (MPR),
- 3-dimensional maximum intensity projection,
- 2-dimensional curved MPR,
- 3-dimensional volume-rendering technique,

Images are often burdened with artifacts, but diagnostic quality may be improved by additional image reconstructions on CT station at alternate times in the cardiac cycle with reduced cardiac motion [14].

(A) (B)

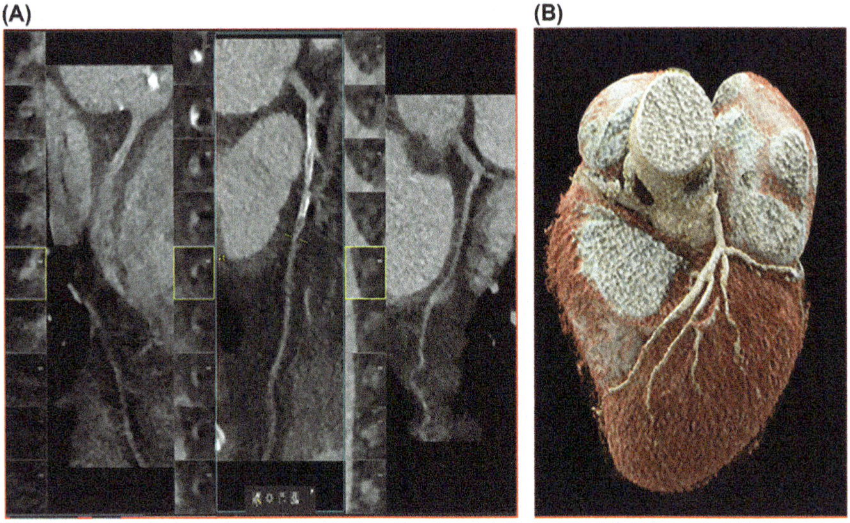

FIGURE 3.8

Artifacts in a cardiac study due to a heart rate of 109 beats per minute. (A) 2D scans of the heart (CT/MRI); (B) a 3D reconstruction of the heart based on the scans presented in (A).

3.4.5 Artifacts

Artifacts may occur during the examination caused by patient movement, ECG recording, and technical problems (Figs. 3.8 and 3.9). The most common are:

- cardiac motion,
- respiratory motion,
- gross patient motion,
- partial volume averaging,
- beam hardening,
- metal or streak artifacts,

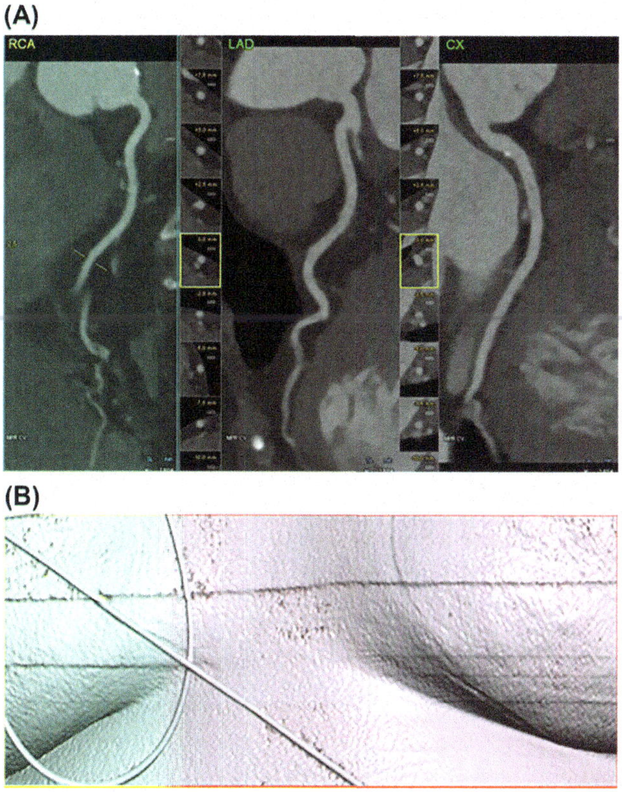

FIGURE 3.9

Artifats in cardiac study due to respiratory motion. (A) medical imaging scans displaying arteries, (B) a chest region with a circular structure and a linear structure intersecting across the vessel, a guidewire or catheter used in a procedure.

- slab or banding artifacts,
- poor contrast enhancement,
- artifacts from overlapping structures.

Deep learning and superresolution reconstructions reduce blooming around calcifications or stent struts and reduce noise, which is especially helpful when the stent diameter is below 3 mm. ECG artifacts can be avoided by performing additional reconstructions at the CT station [15].

3.5 Conclusions

Future directions in CCTA are rapidly advancing, particularly with the integration of artificial intelligence, which enhances image acquisition and postprocessing. Beyond anatomy, emerging tools such as fractional flow reserve derived from CCTA (FFR-CT), perivascular fat attenuation index (FAI), and wall shear stress enable the functional assessment of lesions and provide deeper insight into plaque physiology and vascular inflammation. FAI correlates with coronary inflammation and may predict future adverse cardiovascular events. Dynamic CT is a perfusion-based cardiac functional imaging diagnostic scan with contrast and a hyperemic agent. The result of the examination is displayed as a myocardial perfusion map, showing the extent of ischemic myocardium relative to specific coronary territories, enabling calculation of absolute myocardial blood flow and overcoming the challenges of detecting impaired perfusion in multivessel coronary artery disease [16–18].

References

[1] Li Y, et al. Learning to reconstruct computed tomography (CT) images directly from sinogram data under a variety of data acquisition conditions. IEEE Transactions on Medical Imaging 2019;38(10):2469–81. Available from https://doi.org/10.1109/TMI.2019.2910760.

[2] Ronald B, et al. Technological developments of X-ray computed tomography over half a century: user's influence on protocol optimization. European Journal of Radiology 2020;131:109261. ISSN 0720-048X.

[3] Serruys PW, et al. Coronary computed tomographic angiography for complete assessment of coronary artery disease. Journal of the American College of Cardiology 2021;78:713–36.

[4] Roberts WT, et al. Cardiac CT and CT coronary angiography: technology and application. Heart (British Cardiac Society) 2008;94:781–92. Available from https://doi.org/10.1136/hrt.2007.116392. PMID: 18480352.

[5] Shiyovich A, et al. Photon-counting computed tomography in cardiac imaging. JACC: Cardiovascular Imaging 2026;19(Issue 1):94–117. ISSN 1936-878X.

[6] Kędzierski B, et al. Radiation doses in cardiovascular computed tomography. Life (Chicago, Ill: 1978) 2023;13:990.
[7] Hecht HS. Coronary artery calcium scanning. JACC Cardiovasc. Imaging 2015;8:579–96.
[8] Cheong BYC, et al. Coronary artery calcium scoring: An evidence based guide for primary care physicians. Internal Medicine (Tokyo, Japan) 2021;289:309–24.
[9] Greenland P, et al. ACCF/AHA 2007 clinical expert consensus document on coronary artery calcium scoring by computed tomography in global cardiovascular risk assessment and in evaluation of patients with chest pain: a report of the American College of Cardiology foundation clinical expert consensus task force (ACCF/AHA writing committee to update the 2000 expert consensus document on electron beam computed tomography) developed in collaboration with the Society of Atherosclerosis Imaging and Prevention and the Society of Cardiovascular Computed Tomography. Journal of the American College of Cardiology 2007;49(3):378–402. Available from https://doi.org/10.1016/j.jacc.2006.10.001.
[10] Elkeles RS. Coronary artery calcium and cardiovascular risk in diabetes. Atherosclerosis 2010;210(Issue 2):331–6. ISSN 0021-9150.
[11] He A, et al. Cardiac computed tomography post-transcatheter aortic valve replacement. Cardiovascular Computed Tomography 2024;18:319–26.
[12] Sun Z. Multislice CT angiography in coronary artery disease: technical developments, radiation dose and diagnostic value. World Journal of Cardiology 2010;2:333.
[13] Narula J, et al. SCCT 2021 expert consensus document on coronary computed tomographic angiography: a report of the society of cardiovascular computed tomography. Journal of Cardiovascular Computed Tomography 2020;15:192–217.
[14] Halliburton SS, et al. Do segmented reconstruction algorithms for cardiac multi-slice computed tomography improve image quality? Herz 2003;28:20–31.
[15] Wintersperger BJ, et al. Image quality, motion artifacts, and reconstruction timing of 64-slice coronary computed tomography angiography with 0.33-second rotation speed. Investigative Radiology 2006;41:436–42.
[16] Coerkamp C, et al. Perivascular fat attenuation index (FAI) on computed tomography coronary angiography reclassifies individual cardiovascular risk estimation. International Journl of Cardiology: Cardiovascular Risk Prevention 2025;24:200360.
[17] Antoniades C, et al. Imaging residual inflammatory cardiovascular risk. European Heart Journal 2020;41:748–58.
[18] Rossi A, et al. Dynamic computed tomography myocardial perfusion imaging: comparison of clinical analysis methods for the detection of vessel-specific ischemia. Circulation of Cardiovascular Imaging 2017;10(4):e005505.

CHAPTER

4

The new clinical reality—mixed reality, augmented reality, virtual reality, 3D modeling, 3D printing, artificial intelligence: transformative technologies in medical practice in cardiology

Sandra Zarychta[1,2], Julia Kolasa[1,2], Bernadeta Chyrchel[3,4], Miłosz Rozynek[4,5], Robert Runiński[4,6], Szymon Maciaszek[1,2], Kinga Glądys[1], Michał Karny[1], Robert Żuk[1,2], Piotr Walecki[7], Kaja Tatoń[1,2], Peter Loh[8], Marta Belmonte[9,10], Paweł T. Matusik[11,12], Łukasz Brandt[2,13], Anna Grochowska[5], Grzegorz J. Nalepa[14], Emanuele Barbato[9,10], Tadeusz J. Popiela[5,*] and Klaudia Proniewska-van Dam[1,2,*]

[1]*Center for Digital Medicine and Robotics, Jagiellonian University Medical College, Kraków, Poland,*
[2]*Functional and Virtual Medical 3D Imaging Laboratory, Department of Diagnostic Imaging, University Hospital in Kraków, Kraków, Poland,*
[3]*2nd Department of Cardiology, Jagiellonian University Medical College, Kraków, Poland,*
[4]*Department of Diagnostic Imaging, University Hospital in Kraków, Kraków, Poland,*
[5]*Department of Radiology, Jagiellonian University Medical College, Kraków, Poland,*
[6]*Diagnostic and Interventional Radiology Unit, BIOMORF Department, University Hospital Messina, Messina, Italy,*
[7]*Department of Bioinformatics and Telemedicine, Jagiellonian University Medical College, Kraków, Poland,*
[8]*Department of Cardiology, UMC Utrecht, Utrecht, the Netherlands,*
[9]*Department of Clinical and Molecular Medicine, Sapienza University of Rome, Rome, Italy,*
[10]*Cardiology Unit, Sant'Andrea University Hospital, Rome, Italy,*
[11]*Department of Electrocardiology, Institute of Cardiology, Jagiellonian University Medical College, Kraków, Poland,*
[12]*Department of Electrocardiology, St. John Paul II Hospital, Kraków, Poland,*

Functional and Virtual Medical 3D Imaging in Cardiovascular Medicine. DOI: https://doi.org/10.1016/B978-0-443-44570-5.00002-1

[13]*Department of Electroradiology, Jagiellonian University Medical College, Kraków, Poland,*
[14]*Jagiellonian Human-Centered Artificial Intelligence Department (JAHCAI), Institute of Applied Computer Science, Jagiellonian University, Kraków, Poland*

4.1 Introduction: toward data-driven and three-dimensional-immersive-artificial intelligence cardiology

This chapter presents a dynamically evolving area of modern medicine in which close interdisciplinary collaboration between clinical teams, biomedical engineers, and computer science specialists plays a crucial role. The integration of expertise in clinical practice, medical imaging technologies, and data analysis enables the development and implementation of next-generation tools based on three-dimensional (3D) visualization, biomedical signal processing, and solutions incorporating artificial intelligence (AI) and immersive technologies—virtual reality (VR), augmented reality (AR), and mixed reality (MR), collectively referred to as immersive technologies. These technologies are applied in preoperative planning, medical education, visualization of diagnostic data, and the monitoring of clinical procedures, ultimately improving the safety and precision of therapeutic interventions.

This chapter outlines the broad scope of interdisciplinary team activities—from digital imaging and communications in medicine (DICOM) data segmentation and reconstruction of anatomical structures, through software development and the integration of hospital systems, including Picture Archiving and Communication System (PACS), Radiology Information System (RIS), and Hospital Information System (HIS), to data quality assessment and the creation of modern educational platforms. Particular emphasis is placed on functional imaging and multimodal data integration, which enable simultaneous analysis of anatomy and physiological processes in a spatiotemporal context. This chapter also discusses the limitations of clinical translation of these solutions, including challenges related to interoperability and alignment with existing clinical workflows. The aim of this chapter is to organize the current knowledge in a practical and clinically relevant manner, supporting a clearer understanding of the presented methods and facilitating their effective implementation into routine medical practice.

It is also conceived as a comprehensive, system-oriented overview of the contemporary clinical workflow that transforms raw medical data into immersive, 3D clinical decision support solutions. Drawing on the practical experience of the Laboratory of Functional and Virtual Medical 3D Imaging (3D-FM) from the Department of Diagnostic Imaging, University Hospital in Kraków in Poland, and its interdisciplinary collaborators, this chapter follows the entire translational pathway—from source data acquisition and quality assessment, through multimodal

---

[*]These authors contributed equally and share senior authorship.

integration, segmentation, and 3D reconstruction, to advanced visualization, immersive technologies, AI, and physical model fabrication. Rather than focusing on a single technology, this chapter presents an integrated framework that connects imaging, functional data, computation, and clinical reasoning, highlighting both methodological foundations and real-world implementation challenges. In this way, it aims to provide clinicians, engineers, and researchers with a structured reference that supports informed adoption of immersive 3D technologies in routine clinical practice, education, and procedural planning.

4.2 Interdisciplinary clinical workflow: bridging medicine, engineering, and computer science

The interdisciplinary approach in the work of clinical teams is a unique combination of healthcare facilities with a modern research and technological environment dealing with the development of next-generation 3D medical imaging tools. The main goal of this framework is to develop and implement innovative methods of processing and analysis of medical signals and images, including the use of immersive technologies, functional imaging methods, or elements of AI. These technologies are used in preoperative surgery planning, medical education, visualization of diagnostic data, and monitoring of clinical procedures and follow-up of treatment [1].

The scope of activities of such interdisciplinary teams enables the development and provision of the following data analysis steps:

1. **Creating 3D models**—segmentation of DICOM data, followed by 3D reconstruction of anatomical structures (e.g., heart, blood vessels, and valves), enables precise planning of surgical procedures [2,3].
2. **Use of dedicated software or software development**—using or designing tools and applications to support the analysis and visualization of medical images [4,5].
3. **System integration**—streamlining the communication between PACS, RIS, HIS, and other elements of the hospital infrastructure, for example, electrocardiography (ECG) device for acquiring data on the heart's electrical activation or deploying dedicated medical data analytics software on hospital servers, integrated with hospital information technology (IT) systems [6].
4. **Data quality assessment**—verification of the correctness of medical data, optimization of acquisition protocols, for example, for reconstruction parameters.
5. **Interdisciplinary cooperation**—projects carried out with doctors, IT specialists, engineers, and educators [7].
6. **Exploration of new technologies**—research on the possibilities of using 3D visualization, immersive technologies, and 3D prints in diagnostics and education [7].

7. **Education and data analysis, implementation in clinical workflow**—courses in AI, medical statistics, bioinformatics, and clinical data modeling [7].

3D medical imaging enables precise analysis of anatomical structures and supports diagnostic and therapeutic processes. It encompasses advanced segmentation of DICOM data, performed manually, semiautomatically, or using AI algorithms. A key element of this process is the 3D reconstruction of cross-sectional or volumetric images, most commonly acquired from computed tomography (CT) or magnetic resonance imaging (MRI), which enables detailed representations of body structures. From an imaging perspective, the reliability of 3D reconstructions depends primarily on source data quality. In cardiac CT, ECG-gating strategy, motion control (heart rate/rhythm), contrast timing, and thin-slice reconstructions are key determinants of segmentation accuracy and geometric fidelity. In cardiac MRI, the intended application should guide sequence selection (e.g., cine for function, late gadolinium enhancement (LGE) for scar, mapping/4D-flow for tissue/hemodynamics), as each provides different structural and functional constraints for subsequent modeling. The developed reconstructions can be further displayed as 2D visualizations (on the screen), in a 3D environment (using immersive reality devices), and in the form of physical models obtained using 3D printing.

Immersive visualization technologies using modern devices, such as Microsoft HoloLens 2 headset [8] or Meta Quest 3 as well as 3D prints, are also playing an important role in proposed medical workflow as well as in medical education process [9]. Thanks to these solutions, it is also possible to create interactive anatomical models that are employed in the planning of surgical procedures, which can increase procedural precision and safety and can be used in medical education as a complementary approach [10]. Although these technologies have shown promising results, further research, clinical validation, and large-scale studies are still required before their widespread implementation in routine clinical practice.

4.3 Source data in cardiology: signals, imaging, functional imaging, three-dimensional, and spatiotemporal data representation

Unlike classical structural imaging methods, which focus on mapping anatomy, functional imaging techniques emphasize monitoring and analyzing dynamic physiological processes in real time. 3D functional imaging allows for spatial assessment of biological activity and vital signs, which is used both in diagnostics and therapy planning and monitoring. Among the exemplary solutions we can distinguish the following:

- analysis of signals obtained from various biophysical sensors, such as ECG;
- 3D electrocardiographic imaging (ECGI), which enables spatial reconstruction of the electrical activity of the heart based on signals recorded from the chest surface;

- integration of data obtained from telemetry systems and wearable devices, supporting continuous monitoring of physiological parameters in an outpatient setting;
- modeling biological processes such as hemodynamics based on measurement data and simulation algorithms, enabling the analysis of blood flow and organ function in a spatiotemporal perspective.

With advances in the quality and resolution of medical imaging techniques, 3D organ reconstruction is becoming possible for both clinical and educational purposes. Medical imaging provides numerous two-dimensional (2D) projections (e.g., in DICOM format) of the heart in spatial terms, but their interpretation requires specialist expertise. It is worth noting that, despite the growing interest in and increasing availability of the approaches described above, their clinical translation in cardiology remains constrained by challenges related to interoperability and integration into existing clinical workflows. Multiple studies demonstrate that these technologies are frequently developed as isolated platforms with limited compatibility with existing hospital information systems, including PACS and Electronic Health Records (EHR) [11]. This lack of standardized data exchange impedes the seamless incorporation of these tools into routine diagnostic and interventional pathways, often necessitating manual data handling and parallel workflows. From a workflow perspective, segmentation results should ideally be stored in interoperable formats (e.g., DICOM-SEG where supported) to preserve traceability and enable PACS-centric review. Surface models used for visualization/printing are typically exported as STL/OBJ/PLY; therefore, clear versioning and metadata linkage to the original study are essential for auditability. Given that cardiology workflows are highly time-sensitive and optimized for procedural efficiency and patient safety, technologies that introduce additional operational steps or cognitive burden face substantial barriers to adoption. Consequently, despite increasing evidence of technical feasibility and potential clinical value, many VR-, AR-, MR-, and AI-based solutions remain confined to research, education, or preprocedural planning rather than being fully embedded in real-time clinical care [12,13,14].

In the context of modern medicine and biomedical sciences, the processing of source data and the development of educational competencies are key elements supporting technological and clinical progress. Important areas include the following:

- leveraging advanced data analysis methods, such as statistics, data mining, and AI algorithms to identify patterns, correlations, and predictions from "big data" [15,16];
- effective management of clinical data, including its structuring, anonymization, and ensuring compliance with personal data protection regulations, in particular the General Data Protection Regulation (GDPR [17]);
- interdisciplinary education, focused on the development of practical and theoretical skills among medical and technical professionals, through courses,

workshops, and training programs in the field of data analysis, imaging technology, and modeling of physiological processes;
- creating modern educational materials using immersive technologies (e.g., VR/AR/MR, 3D modeling) and additive technology like 3D prints, supporting the teaching of complex anatomical and clinical issues in an interactive and engaging way [10].

4.4 Multimodal integration of raw medical data for clinical insight

Multimodal cardiac imaging in cardiology is a response to the fact that no single diagnostic technique fully describes the heart—a structure characterized by complex geometry, dynamic function over time, and a strong interdependence between structure and electrophysiology. CT provides high-resolution anatomical information and is particularly useful in assessing valvular structures and vessels; MRI adds morphological and functional assessment with the possibility of tissue characterization; echocardiography enables hemodynamic evaluation; fluoroscopy and C-arm projections form the backbone of real-time guidance during interventional procedures; and electroanatomical mapping and ECG data describe the "electrical dimension" of cardiac activity. In clinical practice, these sources of information often coexist, yet they are frequently interpreted separately, making it difficult to build a coherent picture of the patient. The concept of multimodality assumes that different techniques are complementary: each provides its own "language" for describing the same system, with a distinct balance of strengths, limitations, and potential sources of error. The key value therefore lies not only in access to multiple examinations but also in the ability to combine them within a single spatial context so that anatomical and functional relationships become intuitive and easier to communicate within the team. Integrative platforms such as 3DMedical software (https://iteam.cm-uj.krakow.pl/3d-medical/) [18] within iTeam project (Project number: 2023-1-PL01-KA220- HED-000159314) is developed precisely for this purpose—to support teaching and case analysis by enabling simultaneous visualization of 3D models and imaging data (e.g., CT and DICOM-based X-ray views), as well as linking them to additional modalities. From both an educational and clinical perspective, multimodality has two major dimensions. First, it supports topographical understanding by enabling assessment of spatial relationships between cardiac chambers, valves, the aorta, venous inflows, and adjacent structures—particularly when 2D imaging alone is insufficient. Second, it facilitates connecting anatomy with the procedure, moving from "what do we see?" to "what can we do?"—where procedural planning becomes a model-based process rather than one relying solely on the interpretation of image slices. In the following parts of this subsection, we discuss methods of data integration, the importance of registration and reference landmarks between examinations, and visualization and interaction strategies (including measurements, projections, and data alignment) that transform multimodal imaging into practical support for clinical decision-making. This concept is illustrated in Fig. 4.1, which presents the idea of integrating multimodal imaging (and functional) data to achieve comprehensive case understanding. The integration begins with classical diagnostic modalities: transthoracic echocardiography (TTE) (Fig. 4.1, panel A), providing real-time assessment of cardiac function and hemodynamics; CT (Fig. 4.1, panel B), offering detailed 3D anatomical information; and coronary angiography (Fig. 4.1, panel C), capturing cardiac anatomy and guiding interventional strategies. These datasets can be further enriched by advanced, patient-specific reconstructions: a 3D model of the heart (Fig. 4.1, panel D) augmented with an overlay representing the average position of cardiac electrical activity (pathECG), which links structure with electrophysiological information. Functional electrical data are additionally represented through a conventional ECG (Fig. 4.1, panel E) and its spatially reconstructed counterpart, 3D pathECG (Fig. 4.1, panel F), enabling a more direct correlation between electrical patterns and cardiac anatomy. Finally, the integrated 3D cardiac model can be displayed in MR (Fig. 4.1, panel G), transforming multimodal datasets into an interactive, immersive environment that supports education, collaborative analysis, and procedural planning.

The 3D-FM team developed an interactive, multimodal learning platform, 3Dmedical software [18] (see Fig. 4.2), which uses modern medical imaging technologies, 3D modeling, and tools to visualize physiological functions [19]. By

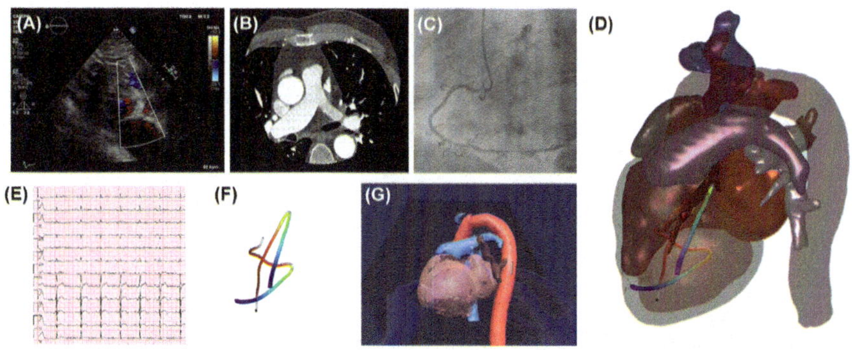

FIGURE 4.1

The idea of multimodal medical imaging (and functional) data integration for comprehensive understanding of the case: Panel A—Transthoracic echocardiogram (TTE); Panel B—Cardiac computed tomography (CT) scan; Panel C—Coronary angiography; Panel D—3D patient-specific model of the heart with overlayed average position of the heart's electrical activity (pathECG) [4]; Panel E—electrocardiogram (ECG); Panel F—3D pathECG; Panel G—3D cardiac model (D) in mixed reality (MR).

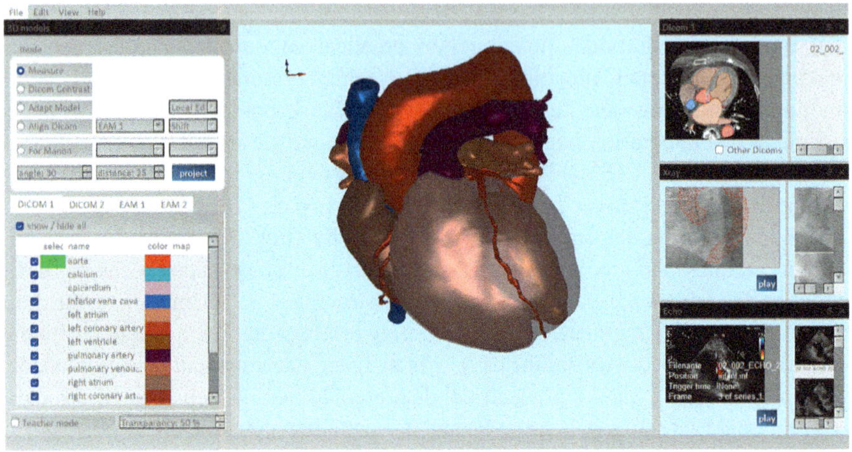

FIGURE 4.2

3DMedical—a multimodal learning platform that integrates medical imaging, 3D models, and tools for visualizing physiological functions, enabling the user to intuitively understand spatial relationships between relevant anatomical structures, their function, procedural planning, and interventional treatment, even in complex cases. The presented example is dedicated to a patient undergoing transcatheter aortic valve implantation (TAVI). Bottom left: a list of segmented structures with opacity and visibility controls; center: an interactive, patient-specific 3D model allowing for exploration of anatomical structures; top right: a cardiac computed tomography (CT) image of the patient's heart with highlighted segmented structures that can be followed by the user to assess, for example, the access path, extent and distribution of calcifications, and can also serve for anatomical measurements; center right: coronary angiography that may assist in correlating in anatomical landmarks with fluoroscopic guidance during the procedure; bottom right: echocardiography, supporting evaluation of valve function and hemodynamics.

integrating data from a variety of sources—such as anatomical imaging, bioelectrical signals (e.g., ECG), 3D models, or functional analyses—it is possible to gain a comprehensive understanding of both the structure and function of the various systems and organs of the human body.

For updates and additional clinical educational cases, see the project website under this reference, where you can find the latest version of the software and clinical educational cases [18,20].

3D visualization allows for the analysis of complex spatial relationships that are often not seen in traditional, 2D imaging. This not only facilitates the teaching process but also supports diagnostics and planning of procedures covering areas from the cardiovascular system through neurology to surgery. The platform offers the possibility of presenting structures and processes in 2D and 3D views, which fosters the integration of knowledge from various fields of medicine.

A multimodal imaging approach—combining anatomical, functional, and bioelectrical data, for example—allows for more precise and personalized patient models. Such solutions, supported by AI and immersive technologies such as AR, VR, or MR (see the model in Fig. 4.3), indicate the direction of development of modern medicine, where education, diagnostics, and treatment are based on the synergy of diverse technologies.

4.5 Three-dimensional reconstruction and modeling of medical imaging data

Following the workflow presented in Fig. 4.4: 3D reconstruction involves converting 2D images, such as CT scans, into a 3D model.

This allows for better imaging and understanding of the internal structures of the patient's body. Such process involves collecting a series of 2D images, which are then—using special algorithms—assembled into a single 3D structure. Interpolation and filtering methods are used to generate a spatial representation. Many systems for medical data acquisition already have built-in algorithms for such a reconstruction.

In the context of medical data visualization, two main techniques are commonly used: volume rendering and surface rendering.

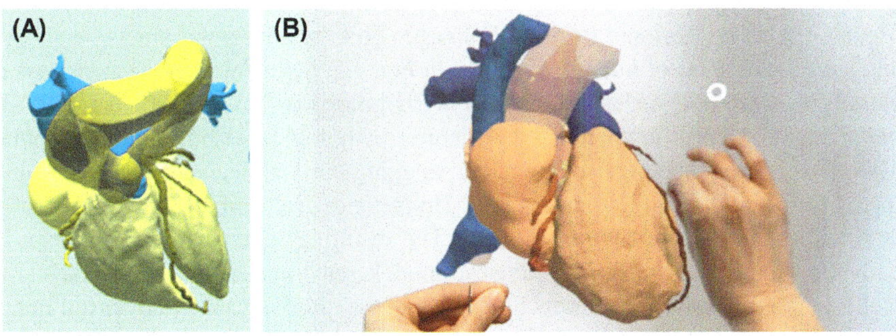

FIGURE 4.3

Learning through immersive, hands-on exploration of a patient-specific 3D cardiac model using the HoloLens 2 mixed-reality headset may support students with a deeper understanding of spatial relationships that are crucial from both an anatomical perspective and for the transcatheter aortic valve implantation (TAVI) procedure. Panel A: adjustment of the appearance of 3D model structures (e.g., color and opacity) in the VSI HoloMedicine—Virtual Surgery Intelligence environment/VSI HoloMedicine (apoQlar medical GmbH, Hamburg, Germany [21]; Panel B: 3D cardiac model of the patient's heart in mixed reality.

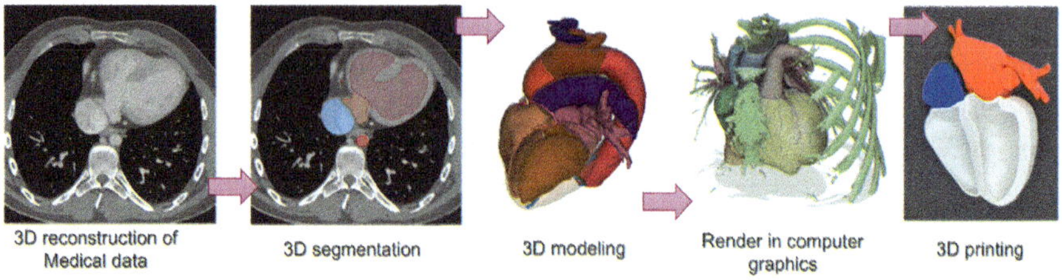

FIGURE 4.4

Workflow for the visualization of multimodal medical raw data, illustrating potential ways in which a patient's data can be used for further analysis, procedural or treatment planning, and the education of medical students [22].

- **Volume rendering** (Volume Ray Casting) allows you to visualize data as a continuous spatial structure or a grid of voxels (3D equivalents of pixels).
- **Surface rendering**, on the other hand, focuses on mapping the outer surfaces of objects using polygonal meshes, typically composed of triangles.

In medical applications, each of these techniques has its own advantages. Volumetric rendering is particularly useful for analyzing CT and MRI data, as it allows for a complete representation of the body's interior—including bones, tissues, organs, and fluids. Surface rendering, on the other hand, is used to create precise models of the outer layers of organs, which is crucial for planning operations.

3D segmentation is used to isolate specific anatomical structures (such as organs, tissues, or lesions) from a 3D image. This facilitates the precise detection and analysis of areas of interest as shown in Fig. 4.5. Segmentation is the process of classifying each image element—pixels in 2D images or voxels in 3D datasets—as belonging to a specific anatomical structure, pathological region, or background. This voxel-level labeling is guided by imaging-derived features such as signal intensity or attenuation values (e.g., Hounsfield Units [HU] in CT), spatial continuity, and anatomical context [23]. The resulting segmented volumes can be converted into 3D surface or volumetric models, enabling quantitative analysis as well as advanced visualization techniques. For this purpose, many advanced image processing techniques are used. Among them are AI algorithms, as well as classic image processing methods such as thresholding or clustering.

3D modeling is the process of creating digital models of objects using specialized computer software. The main goal is to map the geometry of an object into a mesh composed of vertices, edges, and polygons (most commonly triangles). There are many different methods of 3D modeling. Once you've created a geometric model, you can add textures and materials that give it a realistic look, such as color, patterns, or physical properties such as clarity or reflections. These models can be animated and take part in physics simulations, considering gravity, kinematic constraints, collisions, or fluid dynamics.

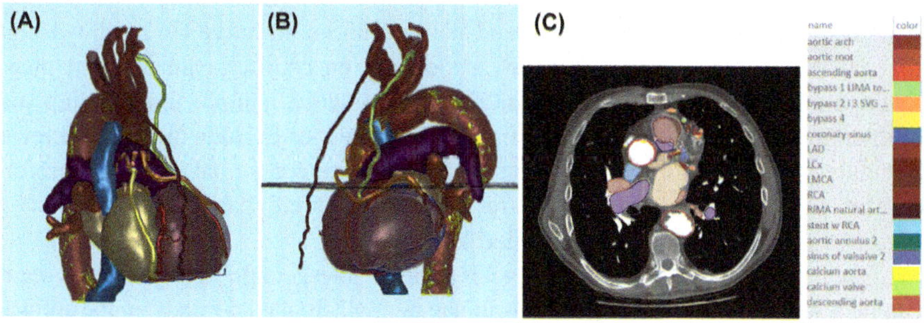

FIGURE 4.5

Different views of the 3D cardiac model in 3DMedical allow the user to explore the heart anatomy of a patient with a history of coronary artery bypass grafting (CABG) and symptomatic severe aortic stenosis undergoing a transcatheter aortic valve implantation (TAVI) procedure: Panel A: Left anterior oblique (LAO) view; Panel B: Right anterior oblique (RAO) view; Panel C: Computed tomography (CT) images with highlighted segmented structures listed on the right.

The final stage of modeling is rendering—that is, generating the final image or animation taking into account lighting, shadows, and other visual effects.

4.6 Segmentation techniques for accurate three-dimensional visualization and preoperative planning

In medical imaging, segmentation refers to delineating anatomical structures or pathological regions within 2D/3D datasets to enable quantitative analysis and the generation of patient-specific 3D models for visualization and procedural planning. In general, methods used for image segmentation can be divided into three main classes: manual, automatic, and semiautomatic.

In the manual approach, the most common tools allow users to "paint" a structure with a brush of a specified diameter, either at a given location or along a desired path, or "draw" on medical image by creating precise outlines (contours), which represent areas composed of individual points along the selected trajectory. Using these methods, the user reconstructs the structure slice by slice, as long as it is visible in the image data. Another technique that can be encountered, based on classical image processing algorithms such as isocontour (iso-value) [24] is level-tracing.† It utilizes the intensity of the seed point and follows surrounding voxels

† In this study, we classify the level-tracing tool as a manual segmentation method in software such as 3D Slicer, since each slice was annotated individually, even though the tool provides semiautomatic contour suggestions.

with similar intensity to trace a closed path around a structure. In cardiovascular CT, HU-based thresholding is highly dependent on contrast enhancement phase, reconstruction kernel, and motion artifacts. In practice, simple thresholding may unintentionally merge chambers and adjacent vessels or exclude clinically relevant structures in low-contrast or complex postoperative/congenital cases. Therefore, careful postprocessing and expert review are required, particularly for valves, coronary arteries, and heavily calcified anatomy.

Among semiautomatic techniques, tools such as "thresholding" or "grow from seeds" can be distinguished. Despite variations in naming, which often depends on the software in which the tool is implemented, the first technique is based on voxel intensity values. The user can define a range (lower and upper thresholds) of HU corresponding to specific anatomical structures, for example, for bone or soft tissue, or even larger areas. These values can be found in the literature or adjusted empirically. Depending on the extent to which the algorithm captures the target structure, postprocessing may be required, and insufficient results may need manual corrections.

Another approach utilizes user-defined "seed" regions placed inside the target anatomical structure and in its background (two separate masks). Based on these seeds, the algorithm grows their respective regions using intensity and spatial information to segment the pointed structure. The resulting segmentation can be refined manually or updated automatically as the seed regions change. While this method reduces manual effort, complex structures often require interaction across multiple slices, and the use of both "main" and nontarget masks can increase the overall segmentation time.

One of the most widely recognized deep learning (DL)-based tools for automatic segmentation of body CT images is TotalSegmentator [25].

Its pretrained models (such as nnU-Net), relying on manually annotated data, allow the user to segment 104 anatomical structures, including organs, bones, and muscles. The model was developed on randomly selected samples from routine clinical studies, including a wide range of patient demographics, pathological conditions, imaging devices, human body structures, acquisition sequences, and locations, enabling robust performance across diverse clinical cases. As reported in the original study, typical failure cases include missing small parts of the colon or iliac arteries and confusing neighboring vertebrae and ribs, despite the high Dice and normalized surface distance (NSD) scores. In 3D-FM team practice, we also observed that the accuracy of resulting segmentations decreases in more complex pathological cases, as well as in arterial or cardiovascular abnormalities.

It is worth noting that whichever method is chosen, results may vary between users and even within the same user over time, which is called inter-user/intra-user variability. Some approaches can reduce, but not eliminate, this difference.

Accurate validation of image segmentation is essential to ensure that its outputs are reliable, reproducible, and clinically meaningful. Validation should not rely on a single metric but instead combine complementary measures, including overlap-based metrics such as the Dice similarity coefficient and boundary-sensitive metrics such as NSD, to capture both volumetric agreement and contour accuracy [26]. This is especially important for small, irregular, or anatomically complex structures, where acceptable volumetric overlap may still conceal clinically significant boundary errors. Segmentation performance must be evaluated on independent, locally representative datasets to confirm robustness across variations in image quality, acquisition protocols, and patient populations. Quantitative evaluation should be supplemented by expert visual assessment to detect clinically relevant errors that may not be adequately reflected by numerical metrics alone.

Moreover, final segmentations intended for clinical decision-making should undergo radiologist expert review, ideally by a radiologist experienced in cardiovascular imaging, with the depth of validation aligned to the intended use (education vs procedural planning vs intra-procedural support). This radiologist-led workflow, supported by the 3D-FM team and routinely applied in our daily practice, ensures traceability, clinical reliability, and high-quality outputs suitable for integration into the clinical pathway. Radiologists play a central and indispensable role in the segmentation process by ensuring that segmentation practices remain clinically grounded and anatomically accurate. Their responsibilities include defining clinically relevant segmentation protocols, standardizing imaging parameters, and supervising segmentation quality across different cases and operators. In AI-assisted segmentation workflows, radiologists provide critical human oversight by reviewing and correcting automated outputs, thereby mitigating automation bias and preventing the propagation of systematic errors into clinical decision-making. These expert corrections also serve as high-quality reference data for iterative algorithm refinement. Furthermore, radiologists contribute to ongoing quality control by jointly assessing technical segmentation performance metrics and workflow-related measures to ensure that efficiency gains do not compromise diagnostic reliability or patient safety [23].

Leveraging the aforementioned processes and the 3D-FM team's experience, we summarized the advantages and disadvantages of each segmentation method in Table 4.1.

In order to better illustrate to the reader the capabilities of the aforementioned approaches, selected segmentation techniques commonly used by the 3D-FM team were visually compared across different levels of anatomical complexity (see Figs. 4.6 and 4.7) using a healthy heart and a congenital heart disease case following the Fontan procedure. Fig. 4.8. illustrates the possible solution proposed by the 3D-FM team that can be helpful in the segmentation process.

As mentioned above, each outcome depends on the input data. Even trained algorithms may sometimes be insufficient for certain cases. This is one of the challenges that the segmentation team frequently encounters. Fig. 4.9. summarizes the most commonly encountered imaging issues and their effects on 3D models.

4.7 Immersive three-dimensional visualization in virtual, augmented, and mixed reality

After a 3D model is prepared, it can be used in a variety of digital applications, such as preparing instructions to 3D-print the model or creating visualizations. The latter

Table 4.1 Summary of strengths and limitations of manual, semiautomatic, and automatic segmentation approaches.

	Manual	Semiautomatic	Automatic
Advantages	**High accuracy** Accurate structure tracing with full user control, even in complex cases or low-contrast images	**Time-efficient** Reduces manual effort while preserving accuracy, especially for large structures	**Fast and scalable** Efficient for large datasets and fast analysis
	Algorithm-independent No limitations imposed by automated methods, even in noisy or irregular data	**User-guided control** Combined algorithmic processing with user input/guidance	**Consistent results** High reproducibility with minimal intra-/inter-user variability
	Universally applicable Suitable for any type of image or structure	**Flexible** Allows for interactive improvement and refinement	**Minimal user effort** Minimal manual correction needed when model performs well, suitable for routine or time-critical applications
Disadvantages	**Time-consuming** Requires significant effort, especially for large datasets	**Input-dependent** Performance depends on quality of initial user input	**Data-quality dependent** Sensitive to noise and acquisition (e.g., low-contrast image) variability
	Fatigue-related error Prolonged manual tasks increase the risk of potential mistakes	**Parameter tuning** Algorithm settings may require tuning across datasets or structures	**Training bias** Potentially reduced generalization to anatomical abnormalities
	Low efficiency Could be not efficient for large datasets or clinical workflows requiring fast results	**Limited in complex cases** Struggles with non- or low-contrast images	**Limited adaptability** Limited to certain structures (e.g., excludes mitral valve)

includes the growing interest in immersive visualizations using MRheadsets, namely, HoloLens 2 or Oculus Meta Quest 3. These emerging technologies still have computing limitations, meaning that when working with surface-rendered

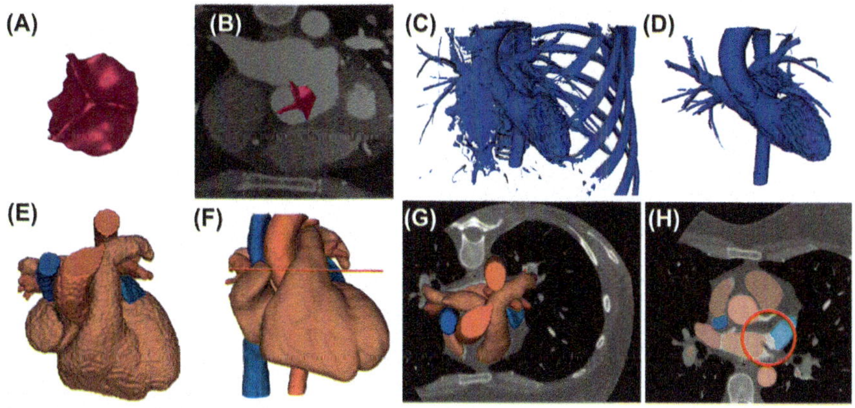

FIGURE 4.6

Healthy heart across different segmentation techniques (results obtained using Slicer 5.8.1, release 2025-03-03). Panels A–B present the manual segmentation approach, which provides highly accurate models and is particularly suitable for small or challenging anatomical structures, such as the aortic valve, where semiautomatic or automatic methods may fail. Panel A shows a 3D model of the aortic valve, while Panel B demonstrates the same model superimposed on the CT image—both views—from 3DMedical software. Despite the precision of manual segmentation, it may become time-consuming and less efficient in clinical workflows that require rapid results, especially for larger structures or low-contrast datasets. Panels C–D illustrate a semiautomatic thresholding approach, which can accelerate segmentation of larger structures but strongly depends on user-defined thresholds and image contrast. The resulting models may range from satisfactory to porous or overly structured representations. In Panel C, only highly contrasted cardiac structures are segmented, and the method does not distinguish morphology, often "stitching" together adjacent anatomical regions such as vertebrae, pulmonary veins, and cardiac chambers. Panel D shows a postprocessed 3D cardiac model that still requires manual correction and expert input to separate structures such as the aorta and left ventricle. Panels E–H demonstrate automatic segmentation, which can save considerable time compared with manual or semiautomatic techniques and provides consistent results; however, manual refinement may still be necessary, particularly in the presence of anatomical abnormalities. Panel E presents the raw TotalSegmentator [25] output, which requires separation of atria and ventricles and does not include coronary arteries or valves. Panel F shows the 3D cardiac model after smoothing; excessive smoothing may reduce clinically relevant geometric detail, including leaflet edges, small calcifications, and coronary lumen dimensions, and therefore smoothing parameters should be selected conservatively and properly documented. Panel G displays the smoothed model overlaid on the corresponding CT image, whereas Panel H highlights structures requiring refinement, such as the left atrium (marked region). Even if extra modules in the software were used, several structures such as the aortic valve are—at this point—are not available in the automatic output and must be segmented manually.

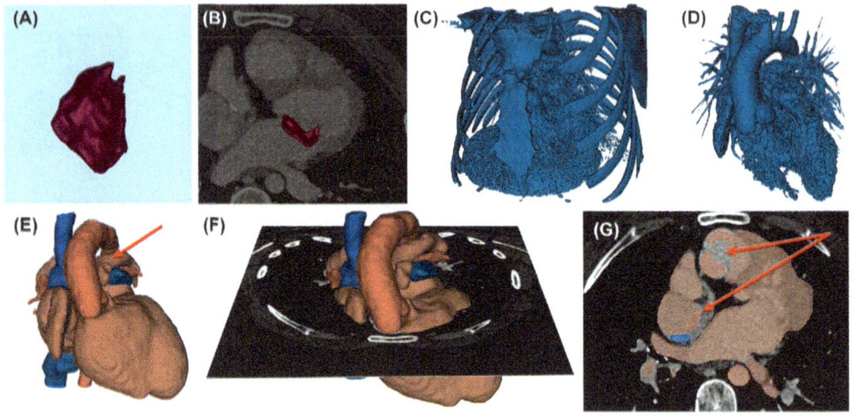

FIGURE 4.7

Congenital heart disease case, following the Fontan procedure, across different segmentation techniques (results obtained using Slicer 5.8.1, release 2025-03-03). Panels A–B: Despite the low-contrast image, including visibility of the mitral valve, an accurate segmentation is still possible without limitations from automated methods. Panel A: 3D model of the mitral valve. Panel B: 3D mitral valve model superimposed on the CT image—both views in 3DMedical software. Panels C–D: Not only the image acquisition but also the complexity of the case itself can affect thresholding results. Hemodynamic changes in the heart after the Fontan procedure result in images with reduced contrast, making the segmentation more challenging and requiring increased manual supervision. Panel C: Segmented cardiac structures are joined with neighboring anatomical regions, such as liver, ribs, and vertebrae; aortic branches (brachiocephalic, left common carotid, and left subclavian arteries) are also segmented, which is not always achieved automatically. Panel D: Postprocessed image still requires manual refinement and tools like smoothing to reduce the porosity of the model. Panels E–G: Automatic segmentation (with the use of TotalSegmentator [25] enables fast delineation of major structures but in this complex case requires careful inspections and corrections. Panel E: Pulmonary artery appears as part of the heart, not a separate structure (red arrow). Panel F: 3D cardiac model (Panel E) sliced by CT and its corresponding image (Panel G) with highlighted structures—missing parts require manual refinement (red arrows).

models, we are constrained by a soft cap for faces—limiting how many triangular faces the model should have.

This is not a hard limit above which triangles would not be rendered. Instead, exceeding the limit will worsen the performance of the MR application, reducing both the usability of the software and readability of the data. For HoloLens 2, it is generally recommended to have less than 100,000 faces per object [27], though complexity of the scene as a whole is also a factor. While this might seem high, a

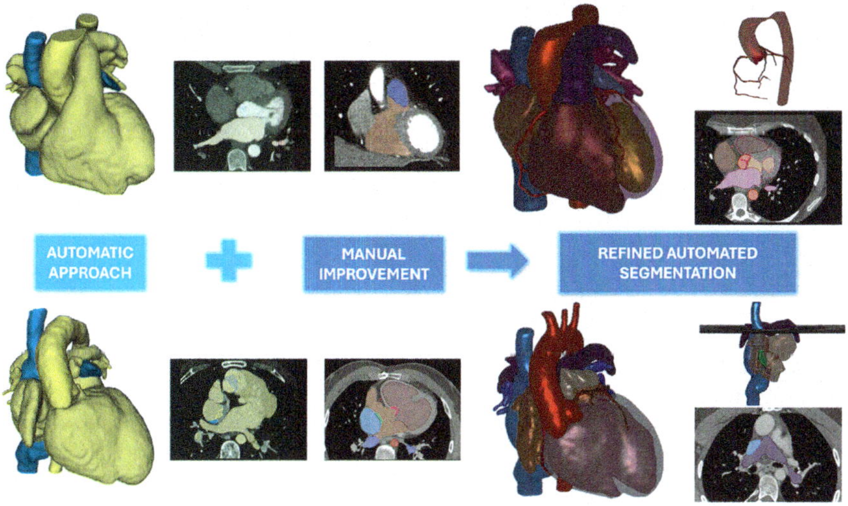

FIGURE 4.8

Refined automated segmentation. Based on the capabilities and visual outcomes of different segmentation techniques, the Laboratory of Functional and Virtual Medical 3D Imaging (3D-FM) team proposed a solution to this challenge. Combined outputs from automatic approach are used as the input data. Manual corrections are then applied (the process can be supported also by semiautomatic tools such as thresholding) to refine the model by correcting regions, improving transitions between anatomical structures, and adding complex or challenging ones. The result is validated by a radiologist, who collaborates on creating the final model. The outcome can be exported in. obj format and integrated with other medical imaging modalities in 3D medical or displayed in mixed-reality environments. Top: Healthy heart. Bottom: Congenital heart disease after Fontan procedure.

completely segmented model of the heart can have over two million faces, requiring significant reduction for satisfying performance.

There are many ways to decimate a 3D model. However, in our approach, it is crucial that the result stays accurate to the original medical data. As such, the prevailing methods revolve around merging small faces or nearby vertices into a single vertex, reducing the details that would've been barely perceptible in the first place. This must be done relative to the size of the object, as small details are much more important for smaller structures such as the coronary arteries or an aortic valve compared to larger structures such as the ventricles or the aorta. Less commonly the model may generate with multiple adjacent faces along the same plane, which can allow for merging them into a single larger face with negligible loss of detail.

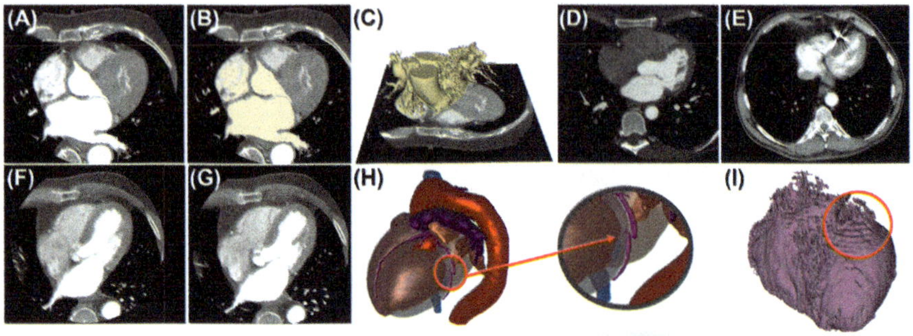

FIGURE 4.9

Medical imaging common issues during segmentation. Panels A–C: Nonuniformly contrasted areas directly impact 3D reconstruction quality and can result, for example, in porous models. Panel A: Raw image; Panel B: Highlighted segmented structures; Panel C: Porosity in the 3D model; Panel D: Low-contrast images impair segmentation, making it difficult to distinguish structures, particularly for automated approaches. In this case, the right ventricle and right atrium are poorly contrasted; therefore automatic methods may struggle to separate them; Panel E: The pacemaker electrodes challenge identification of nearby structures and their contours; Panels G and H: The acquisition process, the breath, motion of the patient, etc., can blur structure borders. Coronary arteries are visible in some slices but may shift between slices. These gaps are clearly seen in the 3D model created based on the same CT scan (Panel J); Panel I—The thickness of the medical image slices can obscure important details, affecting segmentation results.

Having MR-ready models enable us to create our own applications, allowing for more case-specific features. While streamlined solutions such as VSI HoloMedicine offer an easy way to visualize data automatically, it is only the surface of what MR applications are capable of. Finer control over the model and its elements, as well as the possibility for adding additional objects, controls, and functions that work with and around the models themselves, provides an open window into many applications (see Fig. 4.10).

Immersive technologies are revolutionizing the way people interact with digital environments. These solutions enable the creation of uniquely engaging and dynamic experiences by introducing virtual elements into the real environment or completely immersing the user in a simulated space. AI plays a key role in the development of these technologies, making immersive experiences more intelligent, flexible, and tailored to user behavior.

The combination of AI and immersive technologies opens up new possibilities, for example, in the analysis of the anatomically sophisticated structures. It can support precise procedure planning and surgical instruments selection and facilitate

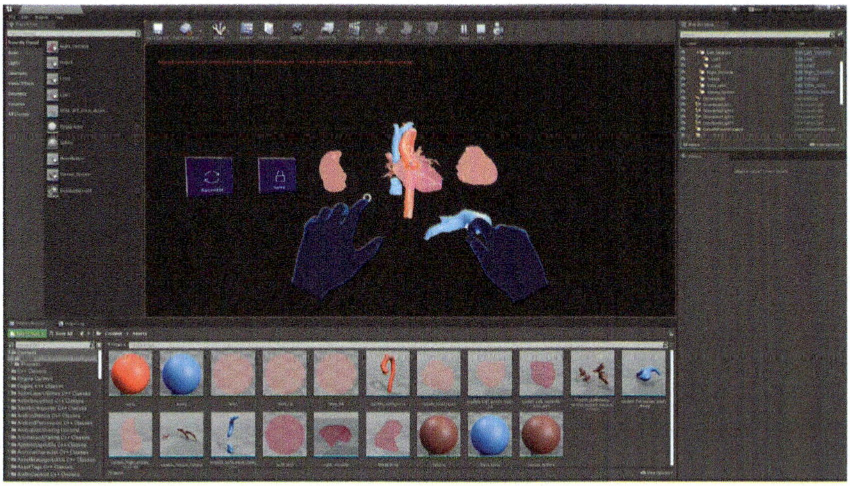

FIGURE 4.10

Screen capture of an MR simulation within Unreal Engine 4.27.2. Unreal Engine allows for full control over MR functionality, from basic material properties to the behavior of individual assets (segmentations) and hand interactions with the model.

intraoperative navigation through real-time visualization, especially in cases involving complex geometry and anatomical interdependencies.

Immersive technologies include both hardware and software components, which collectively allow for the generation of 3D, simulated environments integrated with the physical world. The aim of these solutions is to fully engage the user—to achieve the so-called "immersion" effect. In medical applications, these tools are gaining more and more popularity, especially in areas such as clinical education, diagnostics, rehabilitation, or increasing patient involvement in the treatment process.

Fig. 4.11 shows differences in immersive technology. VR isolates users from the surrounding world, transporting them into a fully digital environment. AR, on the other hand, complements the real image with additional information, such as data visualizations or images. MR combines these two approaches to create a space where it is possible to interact with real and virtual elements at the same time.

In the context of clinical applications, each of these technologies has its own specific function. VR is perfect for operational simulations used in the training of medical professionals. AR can be used to display important data—such as vital signs—in the surgeon's field of view. MR, on the other hand, enables multifaceted collaboration, for example, by real-time joint manipulation of 3D anatomical models.

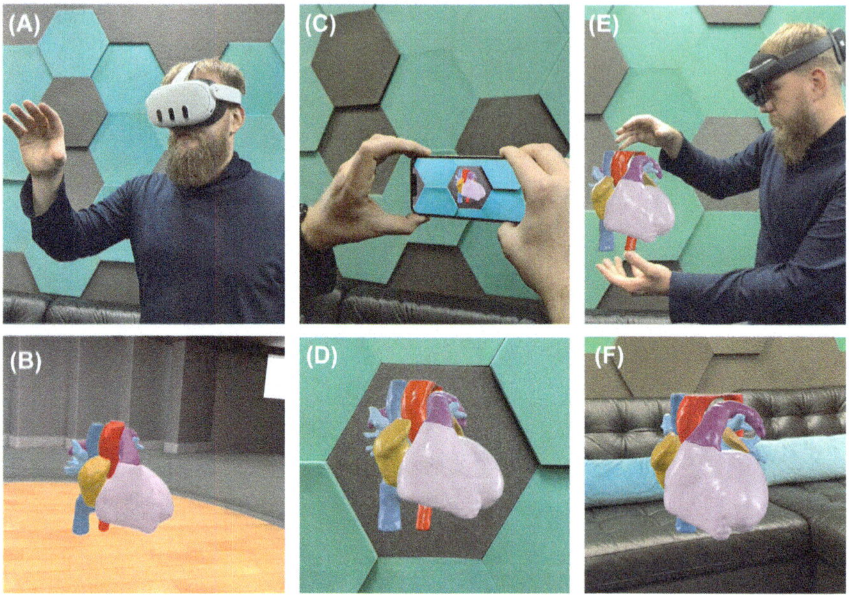

FIGURE 4.11

Immersive technologies provide the user with a unique experience of exploring 3D models, such as cardiac structures, in a more engaging and dynamic way, which is particularly important for educational and clinical purposes. Panels A and B: Virtual reality (VR) using the Oculus Meta Quest 3 headset. Panel A shows a user wearing the VR device, while Panel B presents the corresponding user view in a fully virtual environment, where the anatomical model can be explored and interacted with while the user is isolated from the real surroundings. Panels C and D: Augmented reality (AR) visualization using a smartphone-based application. Panel C shows a user holding a mobile device, while Panel D presents the anatomical model overlaid onto the real environment, displayed according to the application design and implementation. Panels E and F: Mixed reality (MR) visualization using the Microsoft HoloLens 2 headset. Panel E shows a user wearing the MR headset, while Panel F illustrates the corresponding user view, enabling interactive exploration of the anatomical model while maintaining awareness of the surrounding environment.

The integration of immersive technologies with modern imaging methods and patient monitoring systems allows for the development of innovative diagnostic and therapeutic tools. Collecting and analyzing detailed clinical data is crucial—whether in the context of virtual simulations, organ imaging, or long-term health monitoring. As a result, clinicians gain access to high-quality information that supports more accurate diagnoses and more effective treatment.

4.8 From digital models to physical objects: clinical applications of three-dimensional printing

3D printing, commonly referred to as additive manufacturing (AM), is a production methodology based on the layer-by-layer fabrication of physical objects directly from digital 3D model data. In contrast to subtractive or formative manufacturing approaches, AM constructs parts by sequentially joining material according to spatial information derived from computer-aided design files. From a technical perspective, 3D printing involves the interpretation of a digital model—most commonly stored in stereolithography (SLA) (.stl) format—by a printing system that deposits or solidifies material in successive cross-sectional layers. These layers are subsequently bonded or fused, either thermally, chemically, or photochemically, to form a cohesive 3D structure [28]. The International Organization for Standardization defines 3D printing as the fabrication of objects through the controlled deposition of material using a print head, nozzle, or equivalent printing mechanism [29].

In the medical domain, the significance of 3D printing arises from its ability to produce highly individualized anatomical models, patient-specific surgical guides, medical devices, and implants. This level of customization has created new opportunities for improving diagnostic accuracy, procedural planning, and therapeutic outcomes, thereby supporting more personalized and precise patient care [30].

The conceptual foundations of 3D printing can be traced back to speculative visions of automated object fabrication. In 1964, Arthur C. Clarke described a hypothetical "replicator" capable of producing physical objects in a manner analogous to printing written text, anticipating the core idea of AM long before its technical realization. The first concrete technological steps toward modern 3D printing were taken in the early 1980s. Hideo Kodama was among the earliest researchers to propose and experimentally demonstrate rapid prototyping techniques based on photopolymerization. In his pioneering work published in 1980 and 1981, Kodama described a method for the automated fabrication of 3D objects using ultraviolet (UV) light and photosensitive resin. His approach relied on selectively exposing thin layers of photopolymer through a masking system, resulting in the solidification of successive layers and the gradual formation of a 3D object. This concept established the fundamental principles of layer-by-layer fabrication and directly laid the groundwork for what later became known as SLA [31].

Despite the widespread adoption of 3D printing, the classification of AM technologies remains heterogeneous. Different sources propose varying taxonomies, terminology is not fully standardized, and the list of available techniques continues to expand as new methods are developed. Nevertheless, a commonly accepted framework—adopted, among others, by ISO—distinguishes seven principal categories of 3D printing technologies based on the underlying material processing and energy delivery mechanisms (Table 4.2).

Table 4.2 Categories of 3D-printing, description, subcategories [32,29,33,34,35].

Categories of 3D-printing	Description	Subcategories
Vat photopolymerization	Liquid photopolymer in a vat is selectively cured by light-activated polymerization	Stereolithography (SLA) Digital light processing (DLP) Continuous liquid light processing (CLLP)
Powder bed fusion	Thermal energy selectively fuses regions of a powder bed	Selective heat sintering (SHS) Selective laser melting (SLM) Electron beam melting (EBM) Selective laser sintering (SLS) Direct metal laser sintering (DMLS)
Material jetting	Droplets of feedstock material are selectively deposited	Drop-on demand (DOD) Color jet printing Nano particle jetting (NPJ) Continuous inkjet printing (CIJ)
Material extrusion	Material is selectively dispensed through a nozzle or orifice	Fused Deposition Modeling (FDM)/fused filament fabrication (FFF) Pellet extrusion Semisolid extrusion Bioprinting[a]
Directed energy deposition (DED)	Focused thermal energy is used to fuse materials by melting as they are being deposited	Powder feeding type Wire feeding type
Sheet lamination	Sheets of material are bonded to form a part	Laminated object manufacturing (LOM)
Binder jetting	Liquid bonding agent is selectively deposited to join powder materials	–

[a] *(90% of bioprinting uses extrusion, but bioprinting is also carried out using inkjet and laser-based methods (SLA/LIFT).)*

One of the most significant advantages of 3D printing in medicine is the possibility of tailoring medical products to the anatomy of an individual patient. Using imaging data acquired from CT or MRI, it is feasible to generate patient-specific anatomical models, implants, prostheses, and surgical guides that accurately reflect individual anatomical variability. Such personalization has been shown to improve procedural precision, facilitate preoperative planning, and contribute to shorter procedure times, which in turn may reduce the risk of perioperative and postoperative complications.

Based on the practical experience of the 3D-FM, patient-specific 3D-printed models play a particularly valuable role in the preparation of complex interventional and surgical procedures. The ability to rapidly transform digital reconstructions into tangible, spatially accurate models allows clinicians to better understand complex anatomical relationships that may be difficult to fully appreciate using conventional 2D imaging alone. These physical models enable realistic simulation of procedural steps and device positioning, thereby supporting more informed clinical decision-making. In addition, 3D-printed anatomical models have proven to be highly effective educational tools, both in undergraduate and postgraduate medical education, by facilitating the comprehension of complex anatomical structures and pathological conditions through hands-on interaction (Fig. 4.12).

3D printed model of a cardiac cross-section manufactured using fused deposition modeling (FDM) technology at the 3D-FM laboratory. The model consists of three separately printed components produced in different colors and subsequently assembled into a single structure.

Despite these advantages, several limitations constrain the widespread clinical adoption of 3D printing technologies. A notable technical limitation is the restricted build volume of individual 3D printers, which may necessitate segmentation of larger anatomical structures into multiple parts followed by post-print assembly. This process can introduce additional sources of error and increase preparation time. Although the printing process itself is relatively rapid, the overall turnaround

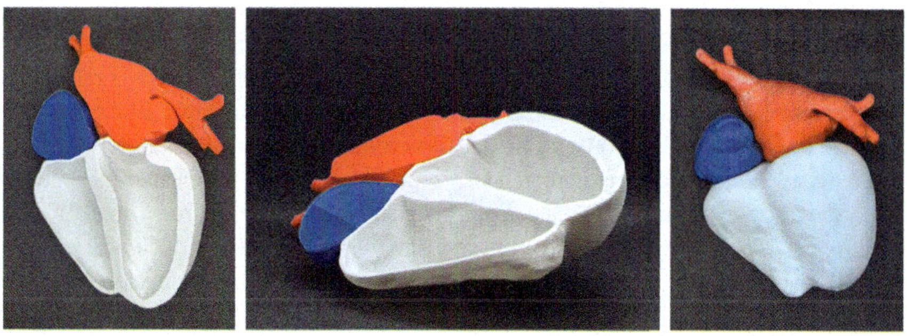

FIGURE 4.12

A 3D-printed cardiac model, including the left and right atria and ventricles, can serve as an effective empirical tool to support both student education and daily clinical practice. The model was fabricated using fused deposition modeling (FDM) 3D printing technology, which allows rapid production of patient-specific heart models and provides a tangible representation of cardiac anatomy for educational, planning, and simulation purposes.

time—including image segmentation, digital model preparation, postprocessing, and sterilization—may be insufficient in urgent or emergency clinical scenarios. From a practical perspective, as observed in routine 3D-FM laboratory workflows, the implementation of 3D printing in clinical practice is also limited by organizational and human resource constraints. The preparation of high-quality, clinically reliable models requires close collaboration between radiologists, engineers, and clinicians, as well as dedicated time for validation and quality control. In many clinical settings, limited staffing and time pressures prevent physicians from actively engaging in the development and operation of 3D printing workflows, which restricts their routine use. Consequently, while 3D printing offers substantial benefits for personalized medicine, its application remains most effective in planned procedures and educational contexts, where sufficient time and resources are available to fully leverage its potential.

4.9 Artificial intelligence in multimodal clinical decision support

Recent years have been marked by significant advances in computational tools, including the development of AI. AI refers to computer systems designed to perform tasks that typically require human intelligence. A key subfield of AI is machine learning (ML), where algorithms learn statistical relationships from data, while DL represents a recent important subset of ML that uses multilayer neural networks to automatically extract complex features, particularly from high-dimensional data such as medical images [36]. In cardiology, AI, especially DL, has shown major impact in cardiovascular imaging by enabling automated image segmentation, quality enhancement, and detection of subtle structural abnormalities [36]. Beyond imaging, AI is increasingly used in cardiology through the integration of multimodal data, enabling improved risk stratification, advanced signal analysis (e.g., ECG), and more accurate outcome prediction. This data-driven approach supports clinical decision-making and personalized treatment planning, allowing for a more comprehensive, precise, and patient-centered management of cardiovascular disease [37]. The main goal is to leverage AI-driven integration of multimodal medical imaging and clinical data to build web-based clinical decision support platforms, including augmented and/or VR solutions, and to develop automated systems for secure access to patients' electronic health records. Building on this approach, the scientific community has increasingly focused on AI-based solutions for medical data analysis, as the medical domain is possibly one of the most sensitive areas of application of AI; an important challenge lies in the interpretability of ML models used.

Commonly, these models might be even hard to understand for specialists, thus contributing to the reduction of trust in the AI solutions. The so-called explainable AI is an active field of research addressing these challenges [38], especially in medicine and healthcare [39].

Medical data is also commonly structured with the use of knowledge graphs, which provide formalized means for organization of conceptual knowledge [40]. Knowledge graphs offer qualitative semantic reasoning compared to the quantitative statistical perspective of ML.

Semantics-aware systems are especially important in the information management, retrieval, and interoperability with EHR [41], as well as in the context of precision medicine [42]. Clinical decision support systems commonly combine versatile AI-based solutions to establish hybrid reasoning frameworks based both on clinical data and domain knowledge [42]. In fact, the trustworthy AI solutions clearly require effective yet responsible combination of AI methods [43]. However, the question of how these methods will handle such data remains an open challenge. By enabling the processing and analysis of vast amounts of medical data in real time, AI is poised to be revolutionary for healthcare systems. As mentioned earlier in cardiology, these rapid advances are already evident across a broad spectrum of applications, ranging from ECG interpretation to clinical decision support systems for interventional cardiology procedures. According to the latest data from the U.S. Food and Drug Administration (FDA) [44], most approved medical devices based on AI and ML are in the field of radiology, followed by the cardiovascular area [44]. The steady increase in the number of such approvals highlights the potential of integrating these technologies into everyday clinical practice. The FDA actively supports this process by developing a transparent and adaptive regulatory framework, such as the Artificial Intelligence and Machine Learning Software as a Medical Device Action Plan, published in 2021 [44].

AI-based technologies enable cardiologists to address the complexity of cardiovascular disease with increasingly sophisticated analytical approaches. These systems facilitate earlier and more accurate diagnosis, refined patient risk stratification prior to targeted interventions, and overall improvements in the quality and consistency of care [45]. In the context of cardiology, AI has the potential to support the entire therapeutic pathway, ranging from preprocedural planning and strategy optimization to enhanced navigation and guidance during interventions, as well as the prediction of perioperative risks and postprocedural outcomes [45].

Given the reliance of cardiology on diverse imaging modalities that provide complementary anatomical and functional information, AI is regarded as a particularly promising technological tool for imaging applications. Advances in AI-driven image reconstruction, analysis, and interpretation are expected to improve the availability, standardization, and quality of cardiovascular imaging data, thereby enabling more robust analytical methodologies and accelerating further developments in precision and personalized cardiology [46].

4.10 Summary

In this chapter, we focused on the dynamically developing area of functional and virtual 3D medical imaging, which is an example of modern, interdisciplinary cooperation between clinical medicine, biomedical engineering, and computer science. We presented how the integration of immersive technologies (VR, AR, and MR), functional imaging methods, data analysis, and AI translates into real benefits in surgical planning, diagnostics, education, and monitoring of patients' conditions. We also discussed the use of 3D reconstruction and modeling of anatomical structures and the use of biophysical data (e.g., ECG), as well as the development of educational platforms based on extended reality (XR) [47]. Recent clinical applications further illustrate how patient-specific cardiac anatomy can be integrated with projected electrical activation within multimodal three-dimensional visualization environments, supporting both clinical interpretation and medical education [48]. We attached particular importance to AI, which enables the analysis of large medical datasets in real time, supports clinical decision-making, and personalizes therapy. Taken together, these developments show how advanced technologies are shaping the future of medicine, making it more precise, safe, and focused on individual patient needs. The importance of the approaches presented in this chapter lies in their ability to transform fragmented clinical information into a coherent, patient-specific representation that supports informed clinical decision-making. Modern cardiology generates vast amounts of heterogeneous data—structural, functional, electrical, and procedural—which, when analyzed in isolation, may limit comprehensive understanding of complex cardiovascular pathology. By systematically integrating raw medical data into unified 3D and immersive environments, clinicians gain an intuitive and spatially accurate overview of anatomy, physiology, and procedural context. This integration enhances preprocedural planning, improves interdisciplinary communication, reduces cognitive load, and supports safer and more predictable interventions. Ultimately, such system-oriented workflows represent a critical step toward precision medicine, where diagnostic insight, therapeutic planning, and education are aligned within a single, clinically meaningful framework.

References

[1] Rudnicka Z, et al. Cardiac healthcare digital twins supported by artificial intelligence-based algorithms and extended reality—a systematic review. Electronics 2024;13(5):866.

[2] Lo Presti G, et al. Assessment of DICOM viewers capable of loading patient-specific 3D models obtained by different segmentation platforms in the operating room. Journal of Digital Imaging the Official Journal of the Society for Computer Applications in Radiology 2015;28(5):518–27.

[3] Kamio T, et al. DICOM segmentation and STL creation for 3D printing: a process and software package comparison for osseous anatomy. 3D Printing in Medicine 2020;6(1):17.

[4] Brigham and Women's Hospital (BWH), H.M.S. 3D slicer. permissive BSD-like (Slicer License)]. Available from: https://www.slicer.org/.

[5] Klaudia Proniewska-van Dam, P.V.D.E.A. 3D medical software, iTeam project. Available from: https://iteam.cm-uj.krakow.pl/.
[6] Bakker AR. HIS, RIS, and PACS. Computerized Medical Imaging and Graphics the Official Journal of the Computerized Medical Imaging Society 1991;15(3):157–60.
[7] Pohlman A, et al. 3-dimensional reconstruction and mixed reality in thoracic surgery: a narrative review and user guide. Journal of Thoracic Disease 2025;17(12):11402–19.
[8] Palumbo A. Microsoft HoloLens 2 in medical and healthcare context: state of the art and future prospects. Sensors (Basel) 2022;22(20).
[9] Zhang W, et al. Immersive virtual reality based on head-mounted display in medical education: a systematic review. BMC Medical Education 2025;25(1):1593.
[10] Proniewska, K., Michał Goncerz, J.D., Malinowski, K.P., Lis, M.,Solewski, B., Piotrowski, M., et al. Transforming anatomy education with mixed reality: a curriculum-based study using a holographic anatomy software suite.
[11] Iqbal AI, et al. Immersive technologies in healthcare: an in-depth exploration of virtual reality and augmented reality in enhancing patient care, medical education, and training paradigms. Journal of Primary Care & Community Health 2024;15. 21501319241293311.
[12] Pérez-Sanpablo AI, Gutiérrez-Martínez JQ-FJ, Lozano- Rodríguez IG, Roldan-Valadez E. Transforming medical imaging: the role of artificial intelligence integration in PACS for enhanced diagnostic accuracy and workflow efficiency. Current Medical Imaging 2025;21.
[13] Faizan Siddiqui M, et al. Integration of Augmented Reality, Virtual Reality, and Extended Reality in Healthcare and Medical Education: A Glimpse into the Emerging Horizon in LMICs-A Systematic Review. Journal of Medical Education and Curricular Development 2025;12. 23821205251342315.
[14] Chen W, et al. High-resolution 3D visualization of human hearts with emphases on the cardiac conduction system components-a new platform for medical education, mix/ virtual reality, computational simulation. Frontiers in Medicine 2025;12:1507005.
[15] Okuyan E. Big data in cardiology. Anatolian Journal of Cardiology 2019;22(Suppl 2):23–4.
[16] Xiao H, et al. Big data, extracting insights, comprehension, and analytics in cardiology: an overview. Journal of Healthcare Engineering 2021;2021:6635463.
[17] Proniewska K, et al. Immersive technologies as a solution for general data protection regulation in Europe and impact on the COVID-19 pandemic. Cardiology Journal 2021;28(1):23–33.
[18] Klaudia Proniewska-van Dam, P.V.D.E.A., iTeam – Interactive teaching of medical 3D cardiac anatomy supported by mixed reality. Erasmus+.
[19] Matusik PT, et al. His bundle pacing in a patient with complete atrioventricular block and congenitally corrected transposition of the great arteries: potential of the use of extended reality and cardiac electrical activity projected into a 3-dimensional heart model. Polish Archives of Internal Medicine 2025;135(1).
[20] Klaudia Proniewska-van Dam, P.V.D.E.A., 3D Medical software.
[21] Avakian A. A comprehensive review of food and drug administration (FDA)-cleared virtual reality technologies in radiology. Cureus 2025;17(6):e85440.
[22] Mitsouras D, et al. Medical 3D printing for the radiologist. Radiographics a Review Publication of the Radiological Society of North America, Inc 2015;35(7):1965–88.

[23] Chupetlovska K, et al. ESR essentials: a step-by-step guide of segmentation for radiologists—practice recommendations by the European Society of Medical Imaging Informatics. European Radiology 2025;35(11):6894–904.
[24] Sharma N, Aggarwal LM. Automated medical image segmentation techniques. Journal of Medical Physics 2010;35(1):3–14.
[25] Wasserthal J, et al. TotalSegmentator: Robust segmentation of 104 anatomic structures in CT images. Radiology: Artificial Intelligence 2023;5(5):e230024.
[26] Klontzas ME, et al. ESR essentials: common performance metrics in AI—practice recommendations by the European Society of Medical Imaging Informatics. European Radiology 2025.
[27] Optimize your 3D objects to use with Dynamics 365 Guides or in mixed-reality components included in apps created with Power Apps.
[28] Niaounakis, M., Biopolymers: processing and products. 2015.
[29] Standardization, I.O.f. ISO/ASTM 52900:2021(en). Additive manufacturing—general principles—fundamentals and vocabulary. Available from: https://www.iso.org/obp/ui/#iso:std:iso-astm:52900:ed-2:v1:en.
[30] Randazzo M, et al. 3D printing in neurosurgery: a systematic review. Surgical Neurology International 2016;7(Suppl 33). S801-s809.
[31] Beaman, J.J., Bawden, J.W.B., Bourell, D.L., Crawford, R.H., Marcus, H.L., McAlea, K.P. Solid freeform fabrication: a new direction in manufacturing.
[32] Habib MA, Elsersawy R, Khondoker MAH. Pellet-based extrusion additive manufacturing of lightweight parts using inflatable hollow extrudates. Journal of Manufacturing and Materials Processing 2025;9(2):37.
[33] Mateti T, et al. An overview of the advances in the 3D printing technology in 3D Printing. In: Pandey JK, editor. 3D printing technology for water treatment applications. Amsterdam: Elsevier; 2023. p. 1–37.
[34] Kumaran M, et al. 6 - Design and topology optimization for additive manufacturing of multilayer (SS316L and AlSi10Mg) piston. In: Salunkhe S, Amancio-Filho ST, Davim JP, editors. Advances in metal additive manufacturing. Cambridge: Woodhead Publishing; 2023. p. 203–17.
[35] Islam A, Rahman MZ. Recent advances in additive manufacturing techniques: an in-depth review. In: Hashmi S, editor. Comprehensive materials processing. Second EditionOxford: Elsevier; 2024. p. 352–78.
[36] Esteva A, et al. A guide to deep learning in healthcare. Nature Medicine 2019;25(1):24–9.
[37] Biondi-Zoccai G, et al. Artificial intelligence in cardiology: general perspectives and focus on interventional cardiology. The Anatolian Journal of Cardiology 2025;29:152–63.
[38] Ali S, et al. Explainable artificial intelligence (XAI): what we know and what is left to attain trustworthy artificial intelligence. Information Fusion 2023;99:101805.
[39] Hossain MI, et al. Explainable AI for medical data: current methods, limitations, and future directions. ACM Computing Surveys 2025;57(6):148. Article.
[40] Tiddi I, Schlobach S. Knowledge graphs as tools for explainable machine learning: a survey. Artificial Intelligence 2022;302(C):21.
[41] Murali L, et al. Towards electronic health record-based medical knowledge graph construction, completion, and applications: a literature study. Journal of Biomedical Informatics 2023;143:104403.

[42] Sutton RT, et al. An overview of clinical decision support systems: benefits, risks, and strategies for success. npj Digital Medicine 2020;3(1):17.
[43] Ahadian P, et al. Ethics of trustworthy AI in healthcare: challenges, principles, and practical pathways. Neurocomputing 2025;661:131942.
[44] FDA. Artificial intelligence in software as a medical device. Available from: https://www.fda.gov/medical-devices/software-medical-device-samd/artificial-intelligence-software-medical-device.
[45] Lüscher TF, et al. Artificial intelligence in cardiovascular medicine: clinical applications. European Heart Journal 2024;45(40):4291–304.
[46] Parminder Singh Manghera, J.L., Artificial intelligence in cardiac imaging: a systematic review of diagnostic, prognostic, and workflow enhancements. 2025.
[47] Jung C, et al. Virtual and augmented reality in cardiovascular care: state-of-the-art and future perspectives. JACC Cardiovasc Imaging 2022;15(3):519–32.
[48] Matusik PT, Zarychta SE, et al. Limitations of classical left ventricular hypertrophy electrocardiographic criteria in hypertrophic cardiomyopathy: Transferring cardiac electrical activity onto a 3D heart model in a patient with recurrent ventricular fibrillations. Polish Heart Journal 2026;1-7. Available from https://doi.org/10.33963/v.phj.112331.

CHAPTER

5 Transformative technologies in teaching cardiovascular medicine

Piotr Walecki[1], Julianna Dąbrowa[2], Sandra Zarychta[2,3], Antoni Cierniak[2], Michał Piotrowski[2], Francesco Burzotta[4], Roberto Ferrari[5], Gianluca Campo[5], Thomas Maal[6], Bo Berends[6], Dylan Duits[6], Nico Bruining[7], Rogier Barendse[7], Bernard Solewski[8], Maciej Lis[8], Michał Goncerz[8], Grzegorz Goncerz[8] and Klaudia Proniewska-van Dam[2,3]

[1]*Department of Bioinformatics and Telemedicine, Jagiellonian University Medical College, Kraków, Poland,*
[2]*Center for Digital Medicine and Robotics, Jagiellonian University Medical College, Kraków, Poland,*
[3]*Functional and Virtual Medical 3D Imaging Laboratory, Department of Diagnostic Imaging, University Hospital in Kraków, Kraków, Poland,*
[4]*Fondazione Policlinico Universitario A. Gemelli IRCCS, Università Cattolica del Sacro Cuore, Rome, Italy,*
[5]*Interventional Cath-Lab, University of Ferrara, Ferrara, Italy,*
[6]*Radboudumc 3D Lab, Radboud University Medical Center, Nijmegen, The Netherlands,*
[7]*Department of Cardiology, Erasmus MC (Erasmus University Medical Center), Rotterdam, Netherlands,*
[8]*Department of Anatomy, Jagiellonian University Medical College, Kraków, Poland*

5.1 Background

The use of immersive technologies, which includes virtual reality (VR), augmented reality (AR), mixed reality (MR), and three-dimensional (3D) imaging, has rapidly increased in the field of cardiovascular medicine. Extended reality (XR) technologies include a combination of VR, AR, and MR technologies. They have enabled clinicians and educators to achieve more advanced, spatially detailed visualization of cardiac anatomy and pathology, surpassing the inherent limitations of conventional two-dimensional (2D) medical imaging [1]. Recently, a review of the applications of VR/MR/XR technologies in cardiovascular medicine indexed by PubMed identified over 2000 studies up to the year 2026. These studies cover areas like medical education, procedural planning, and rehabilitation. This shows an expanding role for immersive technologies in cardiovascular care [2].

Historically, medical visualization in cardiology has progressed from static digital atlases and 2D reconstructions to patient-specific 3D models and, more

Functional and Virtual Medical 3D Imaging in Cardiovascular Medicine. DOI: https://doi.org/10.1016/B978-0-443-44570-5.00006-9

recently, to functional holographic images. These new technologies offer dynamic interaction with cardiac data, including isolating cardiac chambers, valves, and major vessels, simulating blood flow, and overlaying virtual images on real-world anatomy. For example, AR systems can project 3D cardiac structures in real time during procedures, improving spatial understanding and possibly reducing the need for contrast agents and radiation exposure. They also assist in clinical decision-making [3]. At the same time, medical education and clinical training need to adjust to this changing technological landscape. Traditional teaching methods, like textbooks, static drawings, cadaver dissection, and 2D imaging, struggle to convey the spatial and functional complexity of cardiovascular anatomy and procedures. Immersive 3D and XR tools create new teaching methods that can boost learner engagement, improve understanding of spatial anatomy, and help develop procedural skills. For practicing clinicians, using immersive technologies can aid in continuous learning, collaboration among professionals, and better communication with patients. Recent studies highlight the need for structured curricula and solid outcome measures to effectively evaluate and enhance the educational benefits of immersive technologies in cardiovascular medicine [4].

Immersive and volumetric technologies have driven substantial progress in both cardiovascular medicine and medical education. A prominent example is the HoloAnatomy Software Suite, originally developed at Case Western Reserve University (CWRU) in collaboration with the Cleveland Clinic. This platform employs MR technology, using head-mounted displays such as the Microsoft HoloLens to project interactive 3D anatomical holograms [5,6]. Previous studies evaluating HoloAnatomy have demonstrated increased learner engagement and improved knowledge retention compared with conventional anatomy teaching methods [7,8].

Functional and volumetric 3D imaging modalities—including computed tomography (CT), magnetic resonance imaging (MRI), and 3D echocardiography—together with computational modeling approaches such as computational fluid dynamics (CFD) and finite element analysis, have become progressively integrated into cardiovascular clinical workflows. Recent investigations have reviewed the use of advanced cardiac simulation platforms derived from medical imaging data, incorporating DICOM-based segmentation (DICOM PS3.10, NEMA PS3 / ISO 12052), mesh generation, and coupled CFD or fluid–structure interaction modeling, particularly in the context of structural heart disease interventions such as valve repair and left atrial appendage occlusion [9,10].

Interoperability between imaging, simulation, and immersive visualization platforms is an important prerequisite for the successful application of these tools. Established standards, such as DICOM and emerging health data standards such as Fast Healthcare Interoperability Resources (FHIR), are also helpful in the integration of volumetric imaging data with VR and AR. Dedicated interface engines enable the rendering of volumetric cardiac models for interactive educational and clinical use. Typically, this workflow comprises image segmentation, mesh generation, texture mapping, volumetric rendering, and deployment within

MR or VR systems. Recent studies have further demonstrated the feasibility of deep-learning-based mesh extraction from cardiac MRI to generate simulation-ready anatomical models [11].

Accordingly, this chapter presents our experience with functional and virtual 3D imaging in cardiovascular medicine, positioning our work within the context of contemporary evidence. We examine how immersive technologies influence clinical workflows and medical education, and we discuss their opportunities, limitations, and implications for broader adoption in routine cardiovascular care.

5.2 Implementation of extended reality in medical education and practice

The implementation of XR technologies—particularly MR—was conducted within a structured educational framework developed at the Jagiellonian University Medical College (JUMC) as part of the Erasmus+ European project "Mixed Reality Supporting Advanced Medical Education (MRAME)" (Fig. 5.1). This program constitutes a systematic educational implementation model that integrates immersive visualization [12]. This work is currently being continued and further developed within the project "Advanced Medical Education in Extended Reality (edu-XR)," which has received funding under the "Science for Society" program of the Polish Ministry of Science and Higher Education.

FIGURE 5.1

Mixed reality anatomy classes at Jagiellonian University Medical College using the Microsoft HoloLens.

The technological workflow applied in this project was based on an integrated methodological model developed and validated at the JUMC. This approach synthesizes practical experience by combining advanced DICOM-based 3D reconstruction, holographic rendering, and structured immersive education within the domain of cardiovascular medicine. Clinical and anatomical datasets were derived from anonymized cardiac CT angiography, MRI, and 3D echocardiography examinations performed at the University Hospital in Kraków. DICOM datasets were imported into dedicated segmentation platforms, including 3D Slicer, Mimics Innovation Suite, and Philips IntelliSpace Portal. Individual cardiac structures—such as the atria, ventricles, cardiac valves, coronary arteries, and the aortic root—were delineated using manual or semiautomated segmentation techniques, including region-growing and threshold-based algorithms. The resulting segmentation masks were refined to minimize artifacts and ensure anatomical fidelity. Finalized segmentations were exported in STL format to serve as the basis for holographic visualization and simulation (Fig. 5.2).

Surface meshes were further processed using Meshmixer and Blender to generate watertight, topologically consistent models suitable for MR visualization. Mesh optimization steps included smoothing, decimation, and color coding

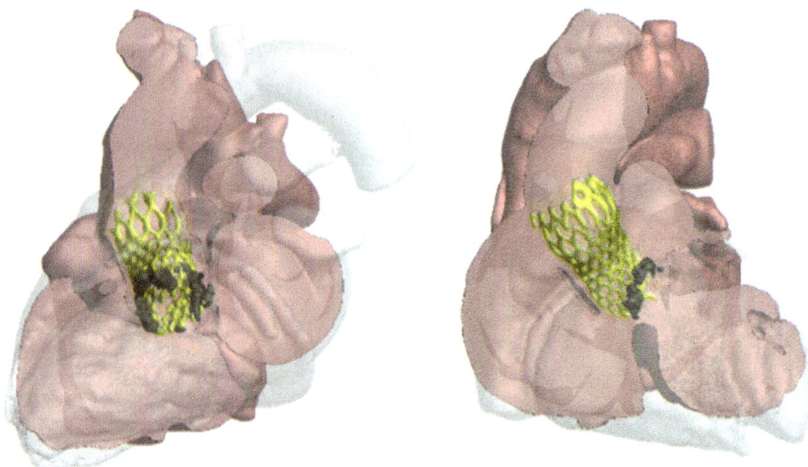

FIGURE 5.2

Heart with a transcatheter aortic valve implantation (TAVI) valve and three-dimensional (3D) model of the prosthesis. (A) Semitransparent 3D reconstruction of the heart with an implanted TAVI prosthesis, showing its position within the native aortic valve and its relation to surrounding cardiac structures. (B) Alternative view of the same model, highlighting the spatial orientation and expansion of the TAVI stent frame (yellow) within the aortic root, enabling detailed assessment of prosthesis placement and anatomical fit in a 3D environment.

according to functional regions of the myocardium and valvular apparatus. In selected clinical cases, CFD preprocessing was performed to simulate intracardiac blood flow and valvular hemodynamics. This component reflected the translational focus of the MRAME framework, in which functional cardiovascular dynamics were visualized alongside anatomical structures to support understanding of complex pathophysiological mechanisms.

The optimized models were subsequently imported into real-time rendering environments based on Unity 3D and Unreal Engine 5, where they were prepared for interactive MR deployment. This stage involved texture mapping, lighting calibration, and implementation of gesture-based user interfaces. Within the HoloAnatomy Software Suite and JUMC-developed holographic modules, the models were deployed on Microsoft HoloLens 2 devices. Users were able to manipulate virtual cardiac models, isolate specific chambers, adjust transparency, and visualize physiological motion synchronized with cardiac cycles. Multiuser synchronization enabled multiple participants to interact concurrently with the same holographic model, thereby supporting collaborative learning and multidisciplinary clinical consultation (Figs. 5.3–5.5).

The implementation incorporated both quantitative and qualitative evaluation components. A cross-sectional survey was designed to evaluate the perception of medical students and academic staff from different European nations regarding MR technology in medical education (Fig. 5.1). The questionnaire was designed to focus on different aspects such as perceived usefulness, learner motivation, cognitive and spatial benefits, health side effects, and feasibility of technology. Quantitative data were combined with thematic analysis to ensure that participants' experiences and contextual information related to immersive learning experiences were considered. A structured educational program focusing on VR, AR, and MR was designed and implemented for medical students. Participants' tolerance to MR technology was evaluated based on different symptoms such as eye strain, sensory overload, and cybersickness. Across all educational interventions, MR technology was found to be effective in learner understanding of spatial knowledge and 3D visualization of anatomical structures, similar to other cross-institutional evaluation studies.

For instance, in educational settings, there was integration of holographic modules in anatomy and cardiology courses that involved preclinical and clinical aspects. Quantitative assessments showed that there was a statistically significant improvement in understanding and retaining knowledge through immersive visualization compared to other conventional methods. In clinical practice, there was use of patient-specific imaging data for creation of holographic reconstructions that aided the Heart Team in preprocedure planning for transcatheter aortic valve implantation (TAVI) and other cardiovascular procedures. The immersive visualization of the aortic root and valvular anatomy aided in multidisciplinary discussion and improved understanding among cardiac surgeons, interventional cardiologists, and imaging experts (Fig. 5.2). These observations illustrate a functional continuum between digital education and clinical application. All educational and evaluation processes were carried out in accordance with the Declaration of Helsinki and were reviewed by the appropriate ethics

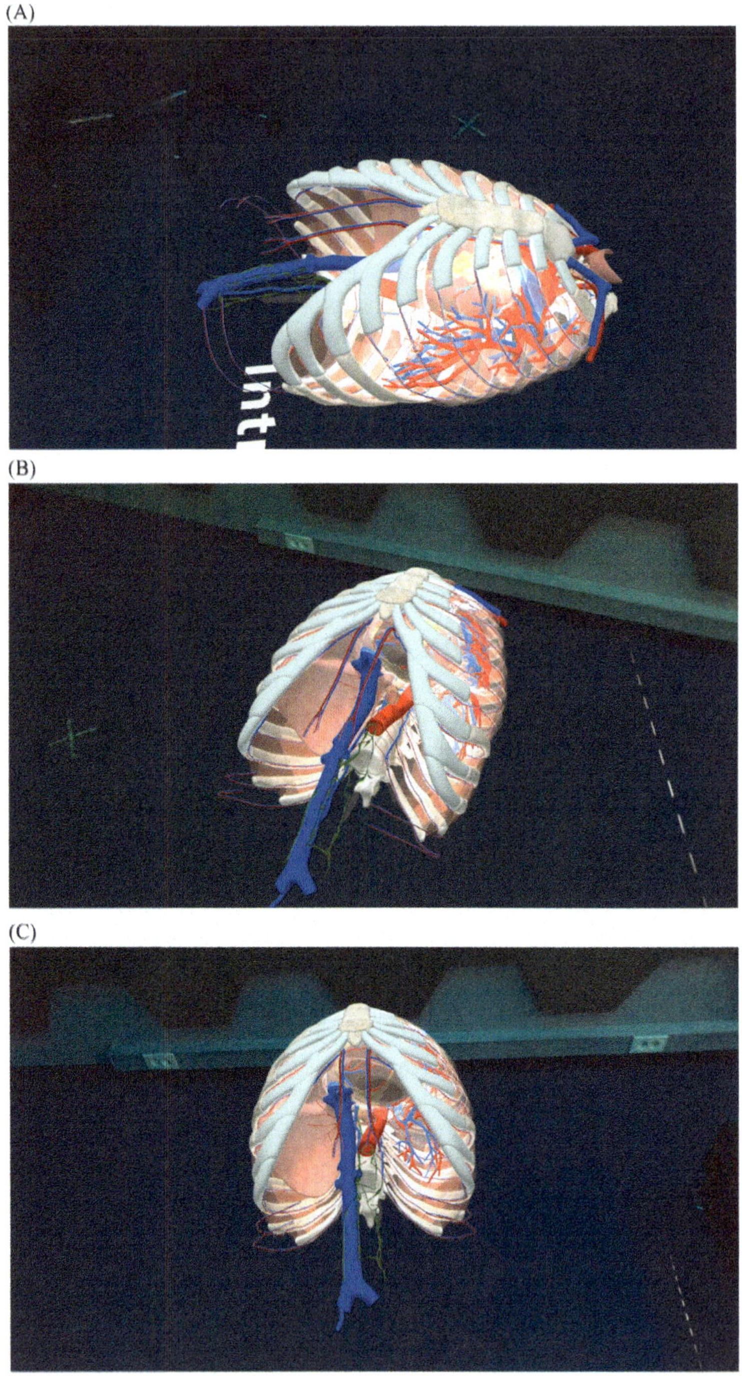

committees. Informed consent was obtained from the participants, and they were advised on the importance of ergonomics and health and safety aspects of immersive technology usage. Technical calibration and monitoring processes were also implemented during the execution. All holographic sessions were preceded by environmental calibration and ergonomic verification procedures. Adverse effects, such as eye strain, dizziness, and sensory overload, were also monitored. However, only a small percentage of the participants experienced mild symptoms, which proved to be transient, thus showing good tolerability of prolonged MR exposure based on the MRAME implementation standards.

This integrative methodological approach—combining educational implementation, mixed-methods evaluation, and translational alignment with clinical practice—establishes a reproducible framework for the introduction of MR technologies in medical education. It positions immersive training as both a pedagogical innovation and a clinically oriented component of digital transformation in contemporary medical curricula.

This technological workflow represents a reproducible workflow for linking medical images, digital reconstruction, and immersive mixed-reality visualizations. The workflow operationalizes the concept of functional digital anatomy, enabling radiological images to be transformed into cognitively accessible holographic models for educational and patient-specific clinical planning activities. The MRAME framework integrates DICOM segmentation, rendering, and pedagogical design to establish a scalable framework for the potential integration of artificial intelligence and digital twin simulation in the field of cardiovascular medicine.

5.3 Educational and cognitive impact of immersive education in cardiovascular medicine

The following findings derive from our original, multicenter study conducted across several European countries, involving more than 600 participants who completed a

FIGURE 5.3

HoloAnatomy lesson—mixed reality (MR) visualization of thoracic anatomy. (A) Oblique view of the thoracic cavity showing the rib cage, sternum, and semitransparent internal structures, including the lungs and major vessels; arteries (red), veins (blue), and selected pathways are visualized to illustrate spatial relationships within the chest. (B) Posterior-oblique perspective highlighting the thoracic skeleton and major mediastinal structures, enabling analysis of the position of the heart, lungs, and great vessels within the rib cage. (C) Frontal view of the thorax demonstrating the anatomical arrangement of the rib cage and internal organs, supporting orientation and understanding of the three-dimensional organization of thoracic structures in an interactive MR environment.

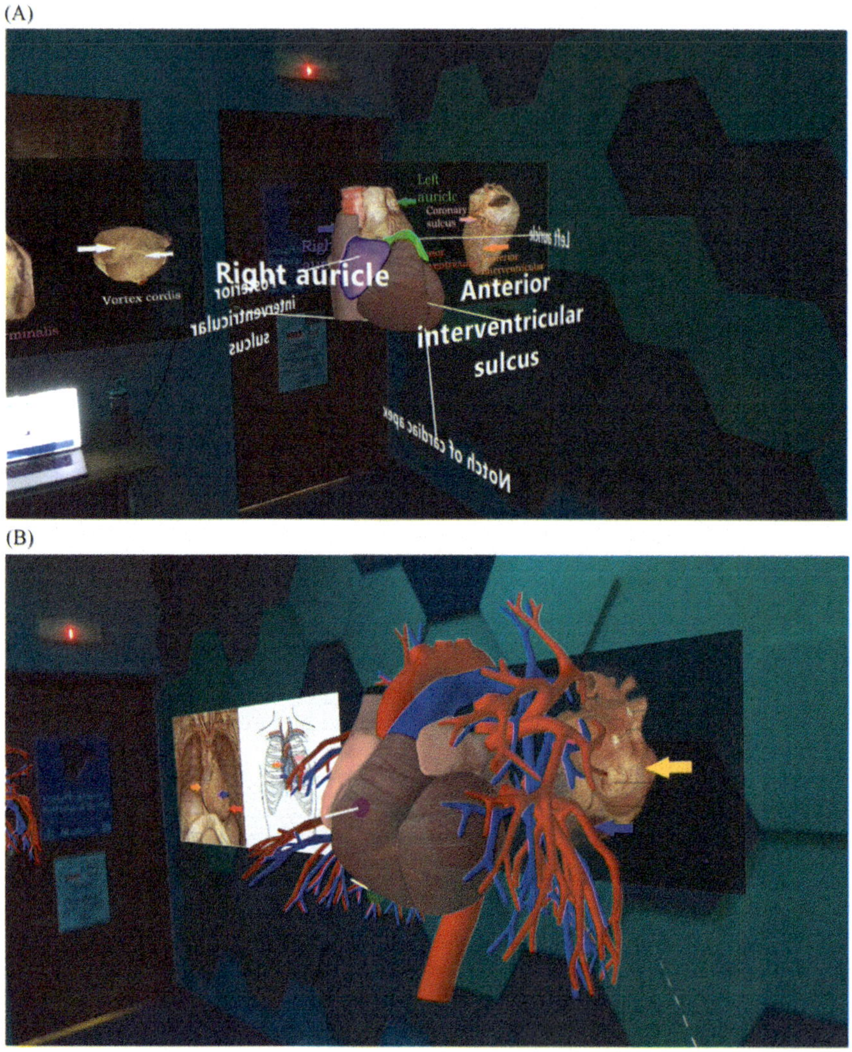

FIGURE 5.4

HoloAnatomy lesson—mixed reality (MR) visualization of cardiac anatomy with integrated educational content. (A) MR scene displaying a 3D heart model with overlaid anatomical labels (e.g., right auricle, interventricular sulcus) and reference images, enabling contextual learning through simultaneous visualization of anatomical structures and supporting educational materials. (B) Interactive 3D visualization of the heart with color-coded vasculature (arterial in red, venous in blue), allowing users to explore spatial relationships between cardiac structures and surrounding vessels within an immersive learning environment.

(A)

(B)

(C)

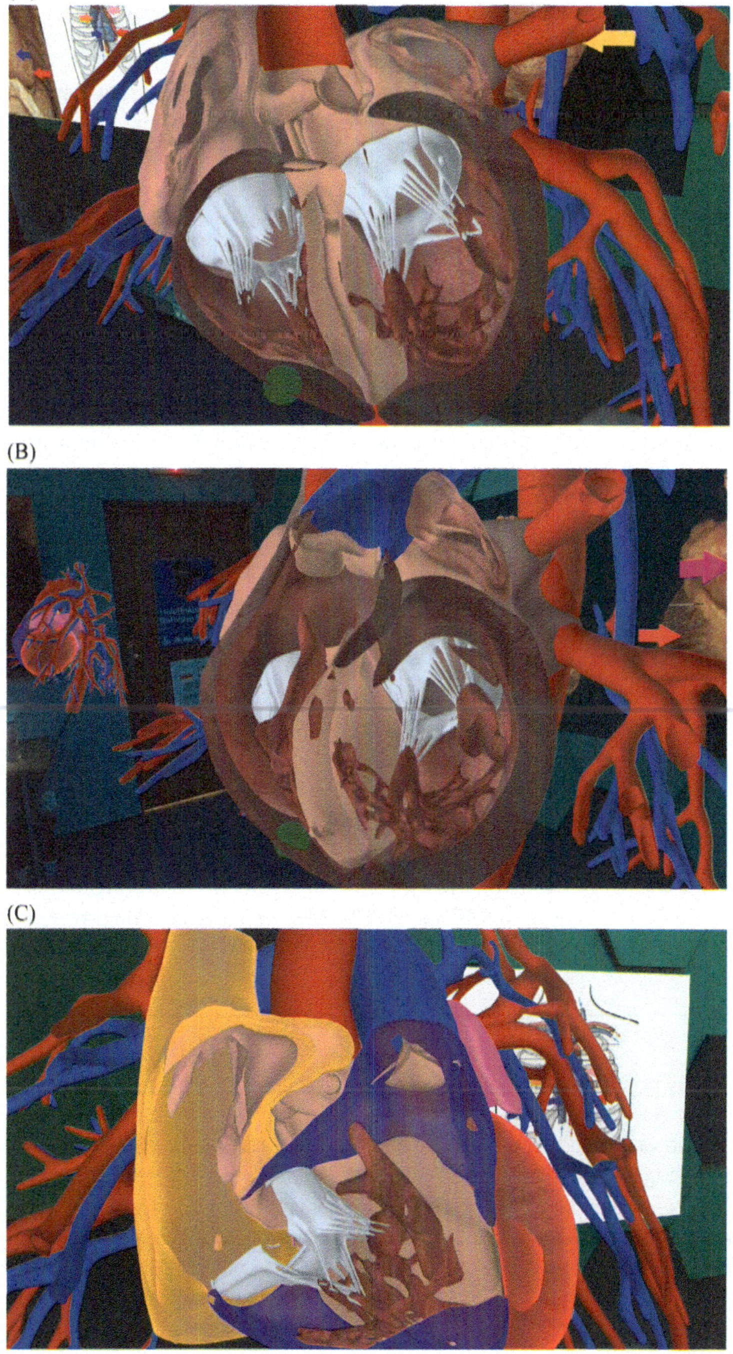

purpose-designed questionnaire developed by the authors (evaluation is ongoing). Quantitative evaluation revealed meaningful differences in perceptions of XR technologies between students and academic educators. Learners consistently perceived XR-based tools as particularly effective for enhancing spatial understanding and clinical visualization, especially in the context of anatomy and cardiology education. Across participating institutions, immersive technologies were broadly regarded as beneficial for improving 3D comprehension of complex anatomical structures and supporting the retention of spatial knowledge.

Academic educators, although recognizing the educational benefits of XR, tended to show more conservative attitudes toward its application, most commonly with regard to cost, access, and integration factors. The analysis of the determinants influencing attitudes toward XR revealed the most significant factors to be professional role and previous experience with immersive technology, whereas demographic factors played a lesser role.

These findings were further substantiated by qualitative thematic analysis. The experience of the student and clinical groups was that they gained an improved understanding of cardiac topography, interchamber relationships, and intracardiac flow patterns through the immersive 3D experience. The MR patient-specific DICOM-based application allowed users to explore the cardiac anatomy from both internal and external perspectives, enabling assessment of the spatial relationship between the aortic root, coronary ostia, pulmonary trunk, atria, and valvular structures. This interaction allows the viewer to achieve a level of spatial understanding and coherence with the cardiac anatomy, which was not possible with the conventional 2D approach. This phenomenon can be explained by the established theories of embodied cognition, which state that the manipulation of 3D as well as holographic images can enhance the memory of the subject for the relevant information, thereby increasing the durability of the learning effects.

Across all educational groups, immersion was associated with increased attention, interest, and self-reported level of engagement, as well as the experience of presence and cognitive involvement. The MR-based educational programs were

FIGURE 5.5

HoloAnatomy lessons—mixed reality (MR) visualization of cardiac anatomy. (A) Internal view of the heart obtained through virtual sectioning, showing the atria and ventricles with clearly visible valves and chordae tendineae; major vessels are presented in contrasting colors to highlight anatomical relationships and blood flow pathways. (B) Oblique sectional view of the heart in an MRmixed reality environment, allowing simultaneous visualization of internal cardiac structures and surrounding vasculature, supporting spatial orientation and understanding of complex anatomy. (C) Color-coded segmentation of cardiac structures, enabling selective isolation and analysis of individual components such as chambers and great vessels, facilitating step-by-step learning in an interactive 3D environment.

found to consistently correlate with better performance in spatial/anatomical reasoning tasks compared to conventional educational methods. Postsession evaluations indicated enhanced accuracy in topographical identification and a strong perception of MR as a valuable adjunct to traditional medical education. Qualitative feedback further suggested improved retention of anatomical relationships and a deeper functional understanding of cardiovascular physiology, including the coordinated dynamics of valvular motion and hemodynamic flow. Participants frequently emphasized that the ability to manipulate holographic models, selectively isolate cardiac chambers, and visualize simulated flow patterns facilitated a more integrated and conceptually coherent understanding of cardiac structure and function.

The integration of immersive technologies in the postgraduate cardiology training programs has further intensified the impact of the aforementioned phenomena. The residents and fellows have claimed that the immersive visualization using MR has helped bridge the cognitive gap between imaging and intraprocedural anatomy. This has helped them make a smooth transition from imaging to execution. The immersive interaction with 3D cardiac models has helped them localize anatomy more precisely and has also enhanced their spatial cognition. The faculty members have also claimed that the immersive experience using holographic technology has helped them engage in active discussions and has transformed passive learning into active learning.

The MR visualization was found to have practical utility in preprocedural planning and communication. In terms of the Heart Team concept, holographic images of the aortic root and valvular anatomical structures were found to aid in mutual understanding among interventional cardiologists, cardiac surgeons, and imaging experts. Patient-specific 3D images obtained by cardiac CT scans enabled real-time discussions of anatomical constraints, access routes, and procedural difficulties. MR visualization was found to aid in consensus building among team members in reducing interpretation ambiguity in the context of preprocedural planning of TAVI and other structural heart interventions. The observational experience of pilot applications indicated better efficiency in planning, particularly in anatomically complex cases, demonstrating the potential of such immersive visualization tools in optimizing team-based clinical decision-making.

These points serve to illustrate the position of immersive XR and MR technologies as clinically enabling tools rather than merely visual aids. Holographic visualization has clear benefits for strengthening interdisciplinary communication and procedural confidence through its ability to translate complex imaging information into intuitive and manipulable 3D representations. Here, immersive technologies can be seen to operate as an enabling technology that bridges diagnostic imaging and clinical decision-making. This serves to reinforce their position within clinical practice and education alike.

The educational and clinical implications of the aforementioned phenomena have shown that XR and MR technologies have the potential to become a significant support system for cardiologists. Immersive visualization has emerged as more than just an educational tool; it has become a clinical tool that can be used for patient-specific procedural planning and has helped in more cohesive and confident decision-making in the field of cardiovascular medicine.

5.4 Stakeholder perspectives and implementation challenges in immersive cardiovascular education

The current results place immersive 3D MR technologies as being highly effective and reproducible in improving spatial understanding, learner engagement, and interdisciplinary communication in cardiovascular-related medical education and clinical practice. When viewed in relation to the current body of literature in anatomy education and cardiology, the results of the current educational program can be seen to support and extend current literature in three main areas: educational effectiveness, clinical decision support, and implementation.

First, the strong acceptance of XR and MR among students, together with observed improvements in spatial comprehension and a clear preference for hybrid educational models, aligns with prior reports on holographic and XR-based anatomy instruction. Previous studies of MR curricula employing platforms such as HoloAnatomy have demonstrated enhanced 3D understanding and learner satisfaction while preserving the role of cadaveric dissection, thereby supporting a complementary rather than substitutive educational paradigm. In this context, our longitudinal, curriculum-wide implementation—combined with evidence of sustained knowledge retention and good tolerability—addresses an important limitation of earlier investigations that were largely confined to short-term interventions or isolated teaching modules [13]. Furthermore, recent educational reviews of XR technologies consistently emphasize that immersive environments can improve visualization and learner engagement, yet highlight the need for structured, curriculum-level evidence supported by mixed-methods evaluation. The present work responds to this need by providing systematically collected observations across multiple cohorts and academic years, offering a more comprehensive perspective on the educational and translational value of immersive technologies in cardiovascular medicine [8,14,15].

Second, the divergence observed between students and educators—whereby students consistently evaluated XR technologies more favorably while academic teachers expressed greater caution—reflects patterns reported across multicountry and multiinstitutional studies on stakeholder acceptance of immersive technologies. Findings from our cross-national investigation corroborate prior evidence indicating that professional role, prior exposure to XR, and selected individual characteristics substantially influence attitudes toward immersive learning environments. Such differences underscore the importance of recognizing heterogeneous stakeholder expectations when implementing XR at scale within medical education. Importantly, despite these differences, an overall positive disposition toward immersive technologies was preserved across both novice and experienced users. High levels of endorsement for XR-based 3D visualization and its potential role in hybrid and remote learning contexts align with recent reports demonstrating a generally optimistic baseline attitude toward immersive technologies, even among users with limited hands-on experience. This suggests that resistance to XR adoption is not rooted in fundamental skepticism toward the technology itself, but rather in pragmatic concerns related to its implementation [16]. These observations support the feasibility of scaling structured XR curricula within medical education, provided that institutional strategies explicitly address educators' concerns. These include considerations of cost-effectiveness, time investment, integration with existing frameworks of assessment, and pedagogical justification. Addressing these factors through faculty development programs, technical infrastructure, and curricular design may prove to be crucial in order to take immersive technologies from the realm of the pilot program to mainstream components of cardiovascular education and training [17,18].

Third, our Heart Team experience with patient-specific holographic reconstructions for TAVI and other structural heart interventions is consistent with emerging cardiology literature that positions XR as a practical, supportive tool for preprocedural planning and intra-team communication. MR platforms have been shown to enhance anatomical understanding of the aortic root, valvular apparatus, and vascular access routes, facilitate collaborative planning, and, in selected reports, influence procedural strategy in a meaningful proportion of cases [19,20]. Systematic reviews and focused analyses published in the *European Heart Journal—Digital Health* and related journals similarly conclude that XR, VR, and AR technologies can support structural heart interventions by integrating multimodal imaging into interactive 3D workspaces, while simultaneously emphasizing the current lack of robust prospective clinical endpoints [3,21]. Our observations—indicating improved clarity of communication, more efficient planning processes, and increased confidence in device sizing and access selection—are aligned with these reports and demonstrate how immersive technologies can be embedded within routine Heart Team workflows rather than remaining confined to proof-of-concept applications [4,22,23]. Recent reviews of AR and MR in cardiology and interventional practice further corroborate that immersive visualization improves spatial precision and may reduce interpretative ambiguity in anatomically complex scenarios, particularly in TAVI, congenital heart disease, and left atrial appendage occlusion. Beyond individual procedures, our findings support the broader conceptual framework in which immersive technologies function as a translational bridge between diagnostic imaging, medical education, and procedural planning in cardiovascular care. XR frameworks have been proposed as integrative layers connecting CT, MRI, 3D echocardiography, and computational modeling with clinician cognition and team-based decision-making [2]. In this context, the combination of curriculum-based MR anatomy education, systematic analysis of XR acceptance among future and current educators, and practical deployment within the Heart Team environment presented in this chapter outlines a coherent translational pathway—from foundational spatial literacy, through advanced interpretation of clinical imaging, to patient-specific interventional planning.

Several limitations and converging concerns must be addressed in the light of the available evidence. The majority of the literature, including our own, has focused primarily on self-report data, short- to mid-term evaluations of knowledge and skill acquisition, and process-related outcomes such as clarity of understanding and efficiency of planning, but not hard endpoints like rates of complications, mortality, and functional outcomes. Systematic reviews of applications of XR technologies in cardiovascular medicine have repeatedly identified this limitation and stressed the need for prospective studies to establish the incremental value of immersive technologies over conventional 3D imaging and high-quality imaging modalities [24]. In parallel, several implementation-related challenges remain only partially resolved. These include the cost of specialized hardware, infrastructure requirements, ongoing software maintenance, workflow standardization, and data protection considerations. Such factors continue to represent practical barriers to broader adoption and are directly relevant to real-world clinical and educational environments [5,25,26].

Despite these constraints, the convergence of our findings with recent reports published in *JACC: Cardiovascular Imaging*, *European Heart Journal—Digital Health* [23], and other high-impact venues supports a clear directional conclusion. Immersive XR technologies, when embedded within structured educational curricula and integrated into multidisciplinary cardiovascular workflows, consistently support improvements in spatial understanding, learner engagement, and team-based communication [3]. By transforming complex imaging datasets into cognitively accessible and interactive representations, these technologies enhance comprehension without displacing the foundational components of medical training and clinical care. The integration of XR into cardiovascular education and practice further reflects a broader pedagogical and technological paradigm shift—from predominantly passive knowledge transmission toward experiential, data-driven, and spatially embodied learning. This shift aligns closely with constructivist learning theories, in which cognitive assimilation is strengthened through active manipulation, exploration, and multisensory interaction. In our work, as in prior studies, the most prominent educational effects were observed in conceptual understanding of spatial relationships rather than in rote factual recall. This, in turn, implies that the role of immersive technologies is to facilitate the more complex cognitive processes, including mental rotation, visual synthesis, and diagnostic reasoning. These observations are consistent with the emerging trend of research indicating that students who are presented with VR and MR experiences demonstrate a more enduring level of retention of anatomical spatial relationships, as well as a more effective transfer of learning to the clinical setting, as opposed to more conventional approaches to instruction. These findings further support the notion that the experience of immersive visualization is not merely a technological enhancement but a true evolutionary step in the manner in which complex cardiovascular information is assimilated, integrated, and applied across the educational and clinical spectrum [27,28].

In cardiovascular medicine in particular, immersive visualization may function as a cognitive scaffold that bridges the gap between diagnostic imaging interpretation and surgical or interventional intuition. Recent studies have demonstrated that holographic 3D reconstructions can improve the efficiency of procedural planning and reduce interobserver variability in preprocedural measurements for TAVI when compared with conventional 2D CT reconstructions [19,29]. Similarly, interactive 3D cardiac models have been shown to enhance diagnostic accuracy in the assessment of congenital heart disease and to streamline case discussions within multidisciplinary Heart Team meetings [30]. These observations are consistent with accumulating evidence indicating that mixed-reality–based Heart Team planning supports more efficient decision-making, reduces ambiguity in device selection, and enhances shared understanding across cardiovascular subspecialties.

Beyond procedural applications, our findings contribute to the broader discussion regarding the translational potential of XR technologies in real-world healthcare systems. The *European Society of Cardiology*'s position statement on digital innovation in cardiovascular education underscores the importance of evaluating XR solutions under pragmatic conditions that reflect heterogeneous institutional infrastructures and resource constraints [31,32]. In this context, the present study demonstrates that immersive modules can be implemented with a relatively low technological burden when data pipelines—such as standardized DICOM-to–mixed-reality conversion and FHIR-based interoperability—are carefully designed. This emphasis on interoperability aligns with prior reports highlighting open data standards as a prerequisite for sustainable integration of XR and artificial intelligence tools into clinical workflows [33]. Nevertheless, caution is warranted to avoid overgeneralization. Not all studies report uniform benefits of XR in medical education or clinical training. Earlier investigations have raised concerns that initial performance gains may be driven in part by novelty effects and could diminish with repeated exposure [34]. In contrast, our longitudinal observations suggest more durable benefits, which may be attributable to the curricular design that positions immersive technologies as a continuous, integrative supplement rather than as a one-time intervention. This distinction highlights the importance of pedagogical structure and longitudinal integration in realizing the full translational value of immersive visualization in cardiovascular medicine.

The implications for faculty development and institutional strategy are equally substantial. Existing evidence indicates that effective implementation of XR technologies is closely linked to senior educators' digital literacy, the presence of institutional incentives, and alignment with established pedagogical and assessment frameworks. For example, a recent case study identified recurrent challenges related to cost, time investment, institutional support, and the absence of standardized workflows for XR deployment [35]. Our findings are consistent with these observations. While educators broadly recognized the transformative potential of mixed-reality technologies, they also identified practical barriers, including technical calibration requirements, hardware logistics, time demands, and integration with existing assessment systems. Overcoming these challenges will require a systemic, institution-wide approach rather than isolated technological initiatives. Such an approach may include the establishment of XR-focused faculty development programs or fellowships, formal integration of MR and XR competencies into accreditation and assessment frameworks, and the creation of shared repositories of validated 3D educational content and modular curricula. These strategies are essential if immersive technologies are to transition from experimental pilot projects to sustainable, scalable components of medical education, rather than remaining isolated innovations with limited long-term impact [36].

Immersive imaging technologies might represent an initial step toward an integrated approach of functional digital twins in cardiology [37]. The combination of holographic visualization, CFD modeling, and patient-specific simulation might represent an initial step toward predictive, personalized, and participatory approaches in cardiovascular medicine. Initial pilot studies evaluating digital heart twins in terms of hemodynamic simulation might indicate an evolution of immersive visualization beyond its educational purpose toward an interactive interface in clinical practice, including therapy planning, scenario evaluation, and even prediction of outcomes [38]. The studies on functional digital twins in cardiology—support the view that immersive visualization technologies are transitioning from experimental to translational maturity [39]. For example, a recent review characterized health-digital twin technologies in cardiology, highlighting XR and artificial intelligence are becoming integral to personalized cardiac modeling [40]. The near-term research agenda should therefore focus on establishing validated outcome measures, cost–benefit analyses, and optimization of human–technology interfaces. Moreover, multiinstitutional randomized studies are required to quantify the additive effect of MR on clinical accuracy and learning retention—especially if XR/MR are to be embedded within structured curricula and multidisciplinary cardiovascular workflows.

Immersive XR and MR technologies hold promise for transforming the field of cardiovascular medicine. Implemented within a structured educational framework, they can lead to improved spatial awareness, engagement, and collaboration among specialists across different disciplines of cardiovascular medicine. However, to move from the experimental phase to the mainstream, further standardization of the technical workflow, faculty, and clinical relevance of XR technologies in cardiovascular medicine is essential. XR technologies appear to be at a crucial juncture, moving from visualization tools to evidence-based, functionally integrated components of digital cardiology education and practice.

5.5 Conclusions

The cumulative body of evidence presented in this chapter, therefore, supports the hypothesis that the aforementioned XR technologies have evolved into powerful tools for the transformation of cardiovascular education and clinical decision-making. The ability of these tools to transform static, 2D data into interactive, spatially coherent holographic images has the power to create a level of conceptual understanding and interprofessional collaboration not previously possible with conventional teaching tools [41]. The educational benefits of the integration of MR technology into the curriculum are significant, as the spatial cognition and procedural understanding of the data were quantitatively improved, with qualitative benefits such as increased learner engagement, collaborative exploration, and increased motivation. However, as with all the aforementioned research, the benefits are also tempered with the limitations and challenges of the technology. The need for specialized hardware, continuous software calibration, and technical supervision are all challenges. Furthermore, some users experienced transient symptoms of cybersickness, visual fatigue, or sensory overload, although these symptoms were not significant enough to deter the benefits of the experience. The aforementioned challenges are similar to the global observations made regarding the use of immersive medical education tools, where users experience transient eye strain or dizziness during initial periods of use [42,43].

In the clinical setting, the utility of 3D XR visualization in the Heart Team setting was evident in the planning of procedures and the promotion of effective communication between cardiovascular specialists. The holographic images of the aortic root, coronary, and valvular anatomy provided a common reference for all, which eliminated ambiguity in interpretation and made the discussion of the procedures more efficient. This is in line with the findings of other studies, which have shown that preprocedure visualization improves the precision of procedure planning and minimizes interobserver variability in complex structural heart procedures [23,24]. At the same time, immersive technologies such as XR and MR should be understood as supplements to, not replacements for, direct clinical experience, tactile training, and conventional procedural workflows.

The principal limitations of our study parallel those noted across the field. The data were predominantly self-reported, relying on subjective perception of utility, comfort, and engagement rather than objective performance or physiological measurements. The participant cohort consisted largely of young learners with limited prior XR exposure, which may have introduced selection bias and restricted generalizability. Moreover, the cross-sectional design and single-institution setting precluded assessment of long-term knowledge retention and the durability of physiological adaptation to immersive exposure. These earlier observations—that many studies of XR/MR in medical education rely on self-evaluated, short-term outcomes and lack true longitudinal follow-up—are supported by findings in the literature [44,45]. Such design limitations underscore the need for future studies to incorporate extended tracking periods, objective outcome measures, and rigorous experimental controls rather than stand-alone pilots.

Future research should therefore adopt longitudinal, multicenter methodologies incorporating both quantitative and physiological endpoints. Objective indicators such as eye-tracking metrics, heart-rate variability, and salivary cortisol should be applied to quantify cognitive load and stress associated with immersive training [42,46]. Expanding participant diversity—across different academic levels, specialties, and prior XR familiarity—will enhance ecological validity. Furthermore, cost-effectiveness analyses and shared infrastructure frameworks are necessary to mitigate financial and logistical barriers[47]. Standardized curricula, controlled exposure durations, and clear pedagogical guidelines must be established to ensure both educational effectiveness and user safety [48].

Not only have immersive technologies advanced beyond experimental usage to achieve clinical and pedagogical validity, but their application is also beneficial in terms of spatial literacy, co-decision-making, and the cultivation of digital competence among future clinicians. However, in order to achieve the full potential of such applications, there is a need to maintain methodological rigor, interoperability standards, and ethical approaches to their application in medicine. In this context, XR in cardiovascular medicine is not just an improvement in visualization; it is an evolutionary shift toward an integrated intelligent model of digital healthcare that is human-centric.

References

[1] Poggianti C, Chessa S, Pelagatti S, Kocian A. Immersive Technologies for inclusive digital education: a systematic survey. Human Behavior and Emerging Technologies 2025;2025:38. Available from https://doi.org/10.1155/hbe2/8888303.

[2] Sun Z, Vaccarezza M. Extended reality and its applications in cardiovascular medicine. Virtual Worlds 2025;4(3):42. Available from https://doi.org/10.3390/virtualworlds4030042.

[3] Kanschik D, Bruno RR, van Genderen ME, Serruys PW, Tsai TY, Kelm M, et al. Extended reality in cardiovascular care: a systematic review. European Heart Journal – Digital Health 2025;6(5):878–87. PMID: 40984992; PMCID: PMC12450517 https://doi.org/10.1093/ehjdh/ztaf070, 40984992.

[4] Moon IT, Ko SK, Kang SH, Yoon CH, Youn TJ, Chae IH. Augmented reality in cardiology: enhancing visualization and precision. Current Cardiovascular Risk Reports 2024;18(12):175–86. Available from https://doi.org/10.1007/s12170-024-00744-7.

[5] Ruthberg JS, Tingle G, Tan L, Ulrey L, Simonson-Shick S, Enterline R, et al. Mixed reality as a time-efficient alternative to cadaveric dissection. Medical Teacher 2020;42(8):896–901. Available from https://doi.org/10.1080/0142159X.2020.1762032. Epub 2020 May 13. PMID: 32401090.

[6] Wish-Baratz S, Gubatina AP, Enterline R, Griswold MA. A new supplement to gross anatomy dissection: HoloAnatomy. Medical Education 2019;53(5):522–3. Available from https://doi.org/10.1111/medu.13845. Epub 2019 Mar 19. PMID: 30891831.

[7] Stojanovska M, Tingle G, Tan L, Ulrey L, Simonson-Shick S, Mlakar J, et al. Mixed reality anatomy using microsoft hololens and cadaveric dissection: a comparative effectiveness study. Medical Science Educator 2019;30(1):173–8. PMID: 34457656; PMCID: PMC8368739 https://doi.org/10.1007/s40670-019-00834-x, 34457656.

[8] Wish-Baratz S, Crofton AR, Gutierrez J, Henninger E, Griswold MA. Assessment of mixed-reality technology use in remote online anatomy education. JAMA Network Open 2020;3(9):e2016271. Available from https://doi.org/10.1001/jamanetworkopen.2020.16271. PMID: 32940677; PMCID: PMC7499123.
[9] Agarwal A, Ranard L, Vahl T, Khalique O. Advanced computer simulation based on cardiac imaging in planning of structural heart disease interventions. Journal of Clinical Medicine 2025;14(19):6885. Available from https://doi.org/10.3390/jcm14196885.
[10] Chessa M, Van De Bruaene A, Farooqi K, Valverde I, Jung C, Votta E, et al. Three-dimensional printing, holograms, computational modelling, and artificial intelligence for adult congenital heart disease care: an exciting future. European Heart Journal 2022;43(28):2672–84. Available from https://doi.org/10.1093/eurheartj/ehac266. PMID: 35608227.
[11] Gaggion N, Matheson BA, Xia Y, Bonazzola R, Ravikumar N, Taylor ZA, et al. Multi-view hybrid graph convolutional network for volume-to-mesh reconstruction in cardiovascular MRI. Medical Image Analysis 2025;104:103630. Available from https://doi.org/10.1016/j.media.2025.103630. Epub 2025 May 17. PMID: 40446480.
[12] Proniewska K, Dąbrowa J, Goncerz M, et al. Transforming anatomy education with mixed reality: a curriculum-based study using a Holographic Anatomy Software Suite. Bio-Algorithms and Med-Systems 2025;21(1):24–39. Available from https://doi.org/10.5604/01.3001.0055.2061.
[13] Mitchell MK, Burns AW, Mohammed S, Balasalle E, Rikard BK, Ghandour S, et al. Augmented reality for anatomy training: efficacy and interest among medical trainees. Journal of Medical Extended Reality 2024;1(1). Available from https://doi.org/10.1089/jmxr.2024.0006.
[14] Ekstrand C, Jamal A, Nguyen R, Kudryk A, Mann J, Mendez I. Immersive and interactive virtual reality to improve learning and retention of neuroanatomy in medical students: a randomized controlled study. CMAJ Open 2018;6(1):E103–9. Available from https://doi.org/10.9778/cmajo.20170110. PMID: 29510979; PMCID: PMC5878950.
[15] Turan Z, Karabey SC. The use of immersive technologies in distance education: a systematic review. Education and Information Technologies 2023;28:1–24. Epub ahead of print. PMID: 37361798; PMCID: PMC10160721 https://doi.org/10.1007/s10639-023-11849-8, 37361798.
[16] Khlaif Z, Salama N, Hamamra B, Mousa A. Factors influencing educators' perspectives on accepting extended reality in health care education: qualitative study. JMIR Medical Education 2025;11:e65042. Available from https://doi.org/10.2196/65042. PMID: 40310667; PMCID: PMC12082055.
[17] Kyaw BM, Saxena N, Posadzki P, Vseteckova J, Nikolaou CK, George PP, et al. Virtual reality for health professions education: systematic review and meta-analysis by the digital health education collaboration. Journal of Medical Internet Research 2019;21(1):e12959. Available from https://doi.org/10.2196/12959. PMID: 30668519; PMCID: PMC6362387.
[18] Kolecki R, Pręgowska A, Dąbrowa J, Skuciński J, Pulanecki T, Walecki P, et al. Assessment of the utility of mixed reality in medical education. Translational Research the Journal of Laboratory and Clinical Medicine 2022;28:100214.
[19] Bonanni M, Russo G, De Siati M, Tomao F, Massaro G, Benedetto D, et al. Holographic mixed reality for planning transcatheter aortic valve replacement.

International Journal of Cardiology 2024;412:132330. Available from https://doi.org/10.1016/j.ijcard.2024.132330. Epub 2024 Jul 2. PMID: 38964558.

[20] Jacquemyn X, Bamps K, Moermans R, Dubois C, Rega F, Verbrugghe P, et al. Augmented and virtual reality imaging for collaborative planning of structural cardiovascular interventions: a proof-of-concept and validation study. Journal of Medical Imaging 2024;11(6):062606. Epub 2024 Oct 8. Erratum in: J Med Imaging (Bellingham). 2024 Nov;11(6):069801. https://doi.org/10.1117/1.JMI.11.6.069801. PMID: 39386378; PMCID: PMC11460359. https://doi.org/10.1117/1.JMI.11.6.062606, 39386378.

[21] Tsai TY, Onuma Y, Złahoda-Huzior A, Kageyama S, Dudek D, Wang Q, et al. Merging virtual and physical experiences: extended realities in cardiovascular medicine. European Heart Journal 2023;44(35):3311–22. Available from https://doi.org/10.1093/eurheartj/ehad352. PMID: 37350487; PMCID: PMC10499546.

[22] Annabestani M, Olyanasab A, Mosadegh B. Application of mixed/augmented reality in interventional cardiology. Journal of Clinical Medicine 2024;13(15):4368. Available from https://doi.org/10.3390/jcm13154368.

[23] Jung C, Wolff G, Wernly B, Bruno RR, Franz M, Schulze PC, et al. Virtual and augmented reality in cardiovascular care. State-of-the-Art and Future Perspectives. JACC Cardiovasc Imaging. 2022;15(3):519–32. Available from https://doi.org/10.1016/j.jcmg.2021.08.017. Epub 2021 Oct 13. PMID: 34656478.

[24] Stephenson N, Pushparajah K, Wheeler G, Deng S, Schnabel JA, Simpson JM. Extended reality for procedural planning and guidance in structural heart disease—a review of the state-of-the-art. The International Journal of Cardiovascular Imaging 2023;39(7):1405–19. Available from https://doi.org/10.1007/s10554-023-02823-z. Epub 2023 Apr 27. PMID: 37103667; PMCID: PMC10250452.

[25] Wickramasinghe N, Thompson BR, Xiao J. The opportunities and challenges of digital anatomy for medical sciences: narrative review. JMIR Medical Education 2022;8(2):e34687. Available from https://doi.org/10.2196/34687.

[26] Chatha WA. From scalpel to simulation: reviewing the future of cadaveric dissection in the upcoming era of virtual and augmented reality and artificial intelligence. Cureus 2024;16(10):e71578. Available from https://doi.org/10.7759/cureus.71578.

[27] Maresky HS, Oikonomou A, Ali I, Ditkofsky N, Pakkal M, Ballyk B. Virtual reality and cardiac anatomy: exploring immersive three-dimensional cardiac imaging, a pilot study in undergraduate medical anatomy education. Clinical Anatomy (New York, N.Y.) 2019;32(2):238–43. Available from https://doi.org/10.1002/ca.23292. Epub 2018 Oct 31. PMID: 30295333.

[28] moro c, birt j, stromberga z, phelps c, clark j, glasziou p, et al. virtual and augmented reality enhancements to medical and science student physiology and anatomy test performance: a systematic review and meta-analysis. Anatomical Sciences Education 2021;14(3):368–76. Epub 2021 Feb 26. PMID: 33378557 https://doi.org/10.1002/ase.2049, 33378557.

[29] Awori J, Friedman SD, Chan T, Howard C, Seslar S, Soriano BD, et al. 3D models improve understanding of congenital heart disease. 3D Printing in Medicine 2021;7(1):26. PMID: 34471999; PMCID: PMC8411549 https://doi.org/10.1186/s41205-021-00115-7, 34471999.

[30] Lee KCS, Breznen B, Ukhova A, Koehler F, Martin SS. Virtual healthcare solutions for cardiac rehabilitation: a literature review. European Heart Journal – Digital Health

2023;4(2):99–111. PMID: 36974268; PMCID: PMC10039430 https://doi.org/10.1093/ehjdh/ztad005, 36974268.

[31] Asselbergs FW, Lüscher TF. Trustworthy implementation of artificial intelligence in cardiology: a roadmap of the European Society of Cardiology. European Heart Journal 2025;46(8):677–9. Available from https://doi.org/10.1093/eurheartj/ehae748. PMID: 39704781.

[32] Samant S, Bakhos JJ, Wu W, Zhao S, Kassab GS, Khan B, et al. Artificial intelligence, computational simulations, and extended reality in cardiovascular interventions. JACC: Cardiovascular Interventions 2023;16(20):2479–97. PMID: 37879802 https://doi.org/10.1016/j.jcin.2023.07.022, 37879802.

[33] Armoundas AA, Ahmad FS, Bennett DA, Chung MK, Davis LL, Dunn J, et al. Data interoperability for ambulatory monitoring of cardiovascular disease: a scientific statement from the American Heart Association. Circulation: Genomic and Precision Medicine 2024;17(3). Epub 2024 May 23. PMID: 38779844; PMCID: PMC11703599 https://doi.org/10.1161/HCG.0000000000000095, 38779844.

[34] Chytas D, Salmas M, Demesticha T, Troupis T. Technological resources in anatomy education: are there any which are worth the effort more than others? Anatomical Sciences Education 2023;16(2):187–8. Epub 2022 Dec 26. PMID: 36524284 https://doi.org/10.1002/ase.2242, 36524284.

[35] Kluge MG, Maltby S, Kuhne C, Evans DJR, Walker FR. Comparing approaches for selection, development, and deployment of extended reality (XR) teaching applications: a case study at The University of Newcastle Australia. Education and Information Technologies 2023;28(4):4531–62. Epub 2022 Oct 20. PMID: 36284824; PMCID: PMC9584278 https://doi.org/10.1007/s10639-022-11364-2, 36284824.

[36] Curran VR, Xu X, Aydin MY, Meruvia-Pastor O. Use of extended reality in medical education: an integrative review. Medical Science Educator 2022;33(1):275–86. PMID: 36569366; PMCID: PMC9761044 https://doi.org/10.1007/s40670-022-01698-4, 36569366.

[37] Sel K, Osman D, Zare F, Masoumi Shahrbabak S, Brattain L, Hahn JO, et al. Building digital twins for cardiovascular health: from principles to clinical impact. Journal of the American Heart Associatio 2024;13(19). Epub 2024 Aug 1. PMID: 39087582; PMCID: PMC11681439 https://doi.org/10.1161/JAHA.123.031981, 39087582. e031981.

[38] Coorey G, Figtree GA, Fletcher DF, Snelson VJ, Vernon ST, Winlaw D, et al. The health digital twin to tackle cardiovascular disease-a review of an emerging interdisciplinary field. NPJ Digital Medicine 2022;5(1):126. PMID: 36028526; PMCID: PMC9418270 https://doi.org/10.1038/s41746-022-00640-7, 36028526.

[39] Rudnicka Z, Proniewska K, Perkins M, Pregowska A. Cardiac healthcare digital twins supported by artificial intelligence-based algorithms and extended reality—a systematic review. Electronics 2024;13:866. Available from https://doi.org/10.3390/electronics13050866.

[40] Katsoulakis E, Wang Q, Wu H, Shahriyari L, Fletcher R, Liu J, et al. Digital twins for health: a scoping review. NPJ Digital Medicine 2024;7(1):77. Available from https://doi.org/10.1038/s41746-024-01073-0. PMID: 38519626; PMCID: PMC10960047.

[41] Tene T, Vique López DF, Valverde Aguirre PE, Orna Puente LM, Vacacela Gomez C. Virtual reality and augmented reality in medical education: an umbrella review. Frontiers in Digital Health 2024;6. PMID: 38550715; PMCID: PMC10973128 https://doi.org/10.3389/fdgth.2024.1365345, 38550715.

[42] Kim HY, Kim EY. Effects of medical education program using virtual reality: a systematic review and meta-analysis. International Journal of Environmental Research and Public Health 2023;20(5):3895. Available from https://doi.org/10.3390/ijerph20053895. PMID: 36900904; PMCID: PMC10001289.

[43] Barteit S, Lanfermann L, Bärnighausen T, Neuhann F, Beiersmann C. Augmented, mixed, and virtual reality-based head-mounted devices for medical education: systematic review. JMIR Serious Games 2021;9(3):e29080. Available from https://doi.org/10.2196/29080. PMID: 34255668; PMCID: PMC8299342.

[44] Walecki P, Dąbrowa J, Więcławek P, Gogól K, Proniewska K. From traditional to immersive: the impact of 3D technologies on medical student learning. Bio-Algorithms and Med-Systems 2024;20(1):90–100. Available from https://doi.org/10.5604/01.3001.0054.9632.

[45] Einloft J, Meyer HL, Bedenbender S, Morgenschweis ML, Ganser A, Russ P, et al. Immersive medical training: a comprehensive longitudinal study of extended reality in emergency scenarios for large student groups. BMC Medical Education 2024;24(1):978. Available from https://doi.org/10.1186/s12909-024-05957-3. PMID: 39252064; PMCID: PMC11385130.

[46] Biniok M, Forbrig TA, Gellert P, Gräske J. Analysis of cybersickness in virtual nursing simulation: a German longitudinal study. BMC Nursing 2024;23(1):187. https://doi.org/10.1186/s12912-024-01833-z. Erratum in: BMC Nurs. 2024 May 28;23(1):355. https://doi.org/10.1186/s12912-024-01971-4. 38509512.

[47] Kantor T, Mahajan P, Murthi S, Stegink C, Brawn B, Varshney A, et al. Role of eXtended reality use in medical imaging interpretation for pre-surgical planning and intraoperative augmentation. Journal of Medical Imaging 2024;11(6):062607. Epub 2024 Dec 5. PMID: 39649776; PMCID: PMC11618384 https://doi.org/10.1117/1.JMI.11.6.062607, 39649776.

[48] Ota Y, Aikawa G, Nishimura A, Kawashima T, Imanaka R, Sakuramoto H. Effects of educational methods using extended reality on pre-registration nursing students' knowledge, skill, confidence, and satisfaction: a systematic review and meta-analysis. Nurse Education Today 2024;141:106313. Available from https://doi.org/10.1016/j.nedt.2024.106313. Epub 2024 Jul 17. PMID: 39042984.

CHAPTER

6 Functional electrocardiographic imaging: advancing cardiovascular diagnostics through innovative imaging and visualization

Peter van Dam[1,2], Jan Kost[1,3], Kinga Glądys[3] and Ksenia Sedova[4]

[1]*Functional and Virtual Medical 3D Imaging Laboratory, Department of Diagnostic Imaging, University Hospital in Kraków, Kraków, Poland,*
[2]*Department of Radiology, Jagiellonian University Medical College, Kraków, Poland,*
[3]*Center for Digital Medicine and Robotics, Jagiellonian University Medical College, Kraków, Poland,*
[4]*Department of Cardiology, Institute for Clinical and Experimental Medicine, Prague, Czech Republic*

6.1 Introduction

The electrical activation of the heart triggers mechanical contraction, followed by active electrical recovery that regulates mechanical relaxation. To ensure the coordinated and autonomous contraction of all myocardial cells, these cells are electrically coupled [1], which results in the electrical activation of all cells once a group of cells is activated. The propagation velocity in myocardial tissue, facilitated by electrical gap junctions, ranges between 0.5 and 0.8 m/s [1].

To initiate the activation of the heart, specialized pacemaker cells and fast-conducting tissue can be found in various parts of the atria and ventricles. The sinus node, located in the roof of the right atrium next to the vena cava superior, consists of pacemaker cells that depolarize autonomously at 60–70 beats per minute or approximately every 900–1000 ms [2]. From the sinus node, which is in the roof of the right and left atria, the atria are simultaneously activated through the atrial myocardial tissue. The electrical activation wave then automatically reaches the triangle of Koch [3], which contains the conduction pathways to the AV node, thereby activating the AV node (see Fig. 6.1) [4,5]. Within this specialized conduction system, activation slowly travels towards the His bundle, which marks the start of the fast-conducting His-Purkinje system. The His bundle quickly divides into left and right branches. The right bundle travels down the right

Functional and Virtual Medical 3D Imaging in Cardiovascular Medicine. DOI: https://doi.org/10.1016/B978-0-443-44570-5.00003-3

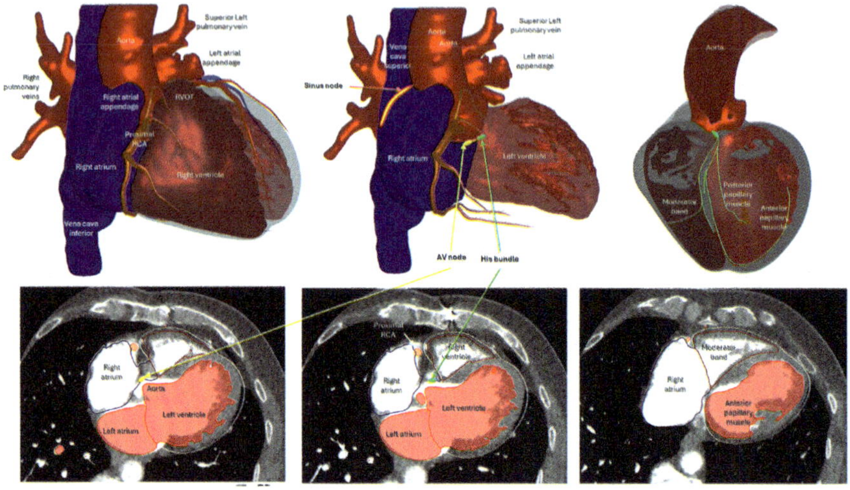

FIGURE 6.1

Anatomical structures of the heart with the relevant electrical structure located in the heart and on the computed tomography (CT) from which the heart was derived. The atrial conductive system, that is, the sinus node and the atrioventricular (AV) node, is indicated in yellow. The ventricular conductive system, that is, the His bundle and His-Purkinje system originating from this node, is indicated in green. Notice that the right His bundle splits into a bundle going over the right ventricle and a bundle on the moderator band, traveling from the right septum to the right ventricular free wall. All atrial and ventricular chambers are indicated both in the three-dimensional (3D) representation as shown in the top panels, as well as in CT cross sections shown in the bottom panels. The contours in the CT images are cross sections of the 3D model. *RCA*, Right coronary artery; *RVOT*, Right ventricular outflow tract.

ventricular septum and splits into a bundle that activates the apical right septum and an arm that travels through the moderator band toward the right free wall. The left bundle further divides into 2–3 bundles that activate the anterior and posterior parts of the left ventricle separately [6]. The His-Purkinje system thus first activates the left ventricle below the His bundle, located near the right coronary ostium of the aorta. The Purkinje network propagates activation at speeds of 2–3 m/s, provided it remains somewhat insulated from the myocardial tissue. In humans, the Purkinje network lies in the lower 70% of the heart and often activates the rest of the ventricles via the papillary muscles of the left ventricle and the connected moderator band of the right ventricle [7].

Electrical activation by the conduction system initiates activation of all connected myocardial cells [8], leading to coordinated contraction of the atrial and ventricular chambers. Afterward, the cells bring the transmembrane potential (TMP) back to its resting state, allowing the cells to relax again. Due to the interconnected nature of myocardial tissue, the ion currents generated by the heart can be measured as potential differences anywhere between two points on or within the body. When measured on the body surface, this is called an electrocardiogram (ECG), and when measured inside the heart, it is known as an electrogram (EGM) (Fig. 6.2). When electrical activity is measured on the heart's surface, the signals show a steep transition from several millivolts (mV) positive to several mV negative, whereas the maximum amplitude on the body surface is usually around 1 mV, with much slower transitions.

With modern electroanatomical mapping systems (EAM), the propagation of the activation can be accurately derived from the EGMs measured on the heart wall (see Chapter 2). These systems are a kind of "three-dimensional (3D) bookkeeping system" in which the relative location within the body is determined by the combination of the strengths of the applied electrical and magnetic fields. These systems have become tools for electrophysiologists in treating cardiac arrhythmias by ablating the origin. Although these systems have made significant improvements over the past decades, several challenges remain [9]. One of the limitations of EAM maps is that only the local activation time on the endocardial or epicardial surface can be measured, from which the 3D propagation has to be derived [10]. Furthermore, these systems are invasive and therefore not applicable for screening or supporting the clinical diagnosis of the ECG. Preferably, cardiac activity would be estimated from ECGs recorded on the body surface. Since the early seventies, scientists, initially primarily physicists, later increasingly computer scientists, have been trying to estimate the electrical activity of the heart using cardiac modeling techniques and mathematical optimization methods to solve the inverse problem in electrocardiography.

In the next paragraph, the specific solution to the forward and inverse problem in electrocardiography will be described, that is, using the equivalent dipole layer (EDL) model, as advocated and refined in the past decades by Adriaan van Oosterom [11]. Despite many recent advances in medical imaging and the reconstruction of patient-specific 3D heart models, this method remains labor-intensive to integrate into the clinical workflow. In the second part of this chapter, the PathECG will be introduced, a method able to estimate the position of average electrical activity over time within a (patient-specific) heart model.

6.2 Forward problem in electrocardiography

The forward problem is the 3D implementation of Ohm's law: U = I x R (see Fig. 6.3). To simulate the ECG and EGM signals, two models are required: a source model (I) and a volume-conductor model (R). The source model represents the electrical sources in myocardial tissue. Such a source model is placed within a realistic volume-conductor model of the torso to enable calculation of the potential anywhere between the source and the body surface. This forward model can then be used in an inverse ECG method, that is, a noninvasive ECG imaging method that tries to characterize the local electrical behavior of the heart from ECG signals measured on the body surface [12–15].

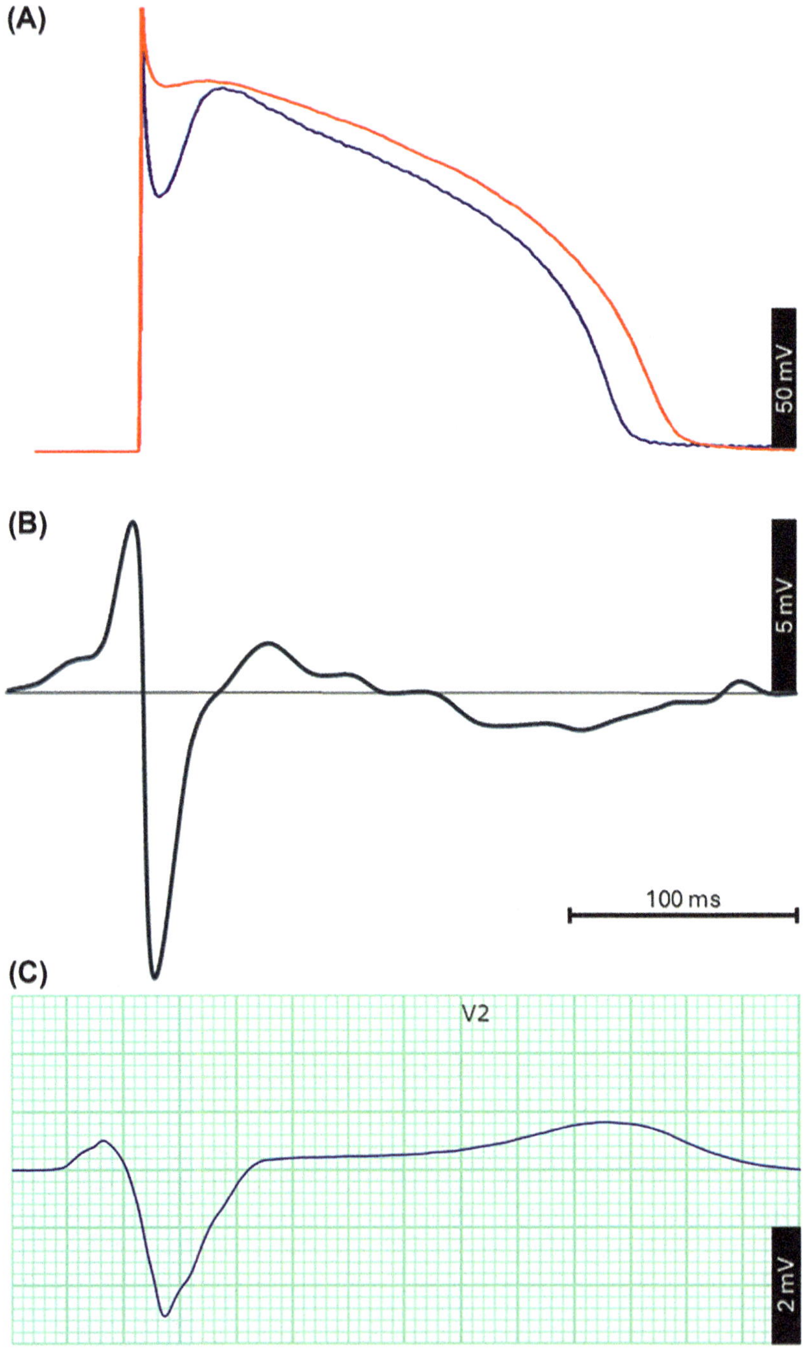
(A)
50 mV
(B)
5 mV
100 ms
(C)
V2
2 mV

6.3 Source models

Any noninvasive imaging method, therefore, needs to postulate a source model. This source model aims to represent the electrical activity of the heart so that it can be used to simulate potentials anywhere in or on the body. The earliest of these is the electric current dipole [16,17], a key element in vectorcardiography (VCG).

From the 1970s onwards, the potential of two other types of source descriptions has been explored. The often-used source model for solving the inverse problem has been the equivalent potential layer model, which estimates the potential distribution on the epicardium of the heart from the body surface potentials [18]. The source model described in this chapter is the EDL model [18]. The advantage of the EDL is that the simulated current leaving the heart is proportional to the local TMP, thus providing a direct link to the electrophysiology of the heart [19–21]. The other used source model to represent the electrical activity is the dipole. Although this source model does not directly link to the TMP, it does link to the heart's anatomy and the direction and position of electrical activity [22].

The macroscopic EDL model [23] is applicable to the entire electrical activity of the atria and ventricles, at any time instant [24]. Several studies [25–27] have shown, using a bidomain model, that the actual current source distribution within the heart is equivalent to a double layer at the myocardial surface, with strength proportional to the local TMP [28]. The waveform of the TMP at each location on the myocardial surface is characterized by two parameters: local activation and recovery times. Consequently, the source parameters consist of the activation and recovery times. The relationship between the source parameters and the source strength is nonlinear; thus a realistic electrophysiological initial estimate is required [12,29,30].

6.4 Volume conductor model

For the EDL source model, we can calculate the infinite medium potential Φ_∞ at any point in an infinite medium. The torso, however, is a bounded medium, and in a

FIGURE 6.2

Examples of different electrical signals measured from the ventricles. (A) Two examples of transmembrane potentials (TMPs) of a cardiac cell (see also Section 6.6). The amplitude is about 100 mV, with the initial jump representing heart activation. Immediately after this jump, early repolarization begins, followed by a plateau phase during which calcium is released and the cell contracts. This is followed by the recovery phase, during which the cell returns to its resting state. (B) shows an electrogram of signals measured directly in the blood cavity of the right ventricle. The amplitude jump is about 13 mV, a significant drop compared to the amplitude jump of the TMP. (C) ECG at V2 as measured on the chest of a normal healthy subject. The amplitude is only 3 mV, again a significant drop in amplitude compared to the local electrogram. This drop is due to the distance from the heart to the V2 electrode. The black bar indicates the amplitude scale. The horizontal line shows the 100-ms duration. Not that similar electrical signals can be obtained from the atria, that is, the P wave.

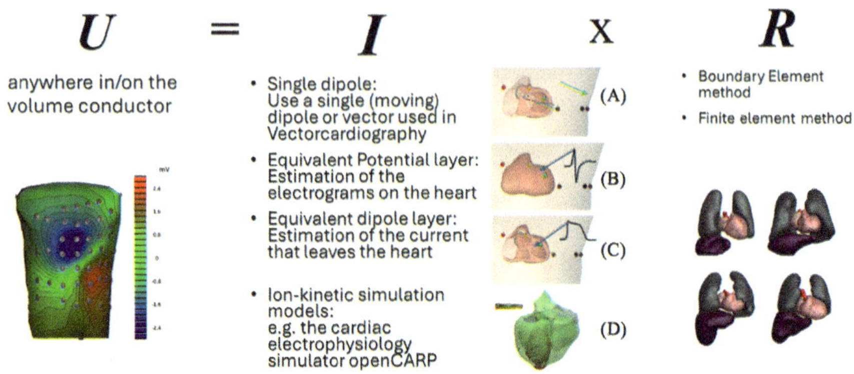

FIGURE 6.3

The forward problem: Ohm's law in 3D. The potentials on the body surface (left panel, with local maxima [red] and local minima [blue]) show the potential distribution. Several equivalent source models exist, of which 4 are shown in this figure: (A) dipole, (B) equivalent potential layer model, (C) equivalent dipole layer model, and (D) and advanced ion-kinetic model, simulating the currents inside and outside the cardiac cells. This latter one often uses a finite element method to compute the potential distribution within the body and on the body surface. The other models can use a boundary element method to compute the potentials arising from the heart.

bounded medium surrounded by an insulator (i.e., the air), the currents cannot leave the body. This results in an adaptation of the potential distribution within the bounded medium, as do the inhomogeneous elements within the torso, for example, blood and lungs [31] (see Fig. 6.4. Differences in individual heart size and the orientations of the heart, lungs, and liver are easily identifiable and thus will influence the solution of the forward model used to simulate the electrical signals. Consequently, the anatomy of the heart (Fig. 6.1) is required to interpret the electrical waveforms arising from the heart.

The volume-conductor model describes the electrical conductivity of a bounded medium in 3D space and how this affects the potential at any observation point. The numerical method used to model the volume conductor is the boundary element method (BEM). The advantage of the BEM over, for instance, the finite element method is the reduction in dimensionality, and therefore, potentials have only to be calculated at the surface boundaries of conductivity transitions. With current computers, however, this reduction in dimensionality is no longer a showstopper. The major inhomogeneous elements within the torso that have to be considered are the lungs and the blood in the heart. The conductivity of the other tissues is about the same as the conductivity of the heart tissue itself. To calculate the potential at

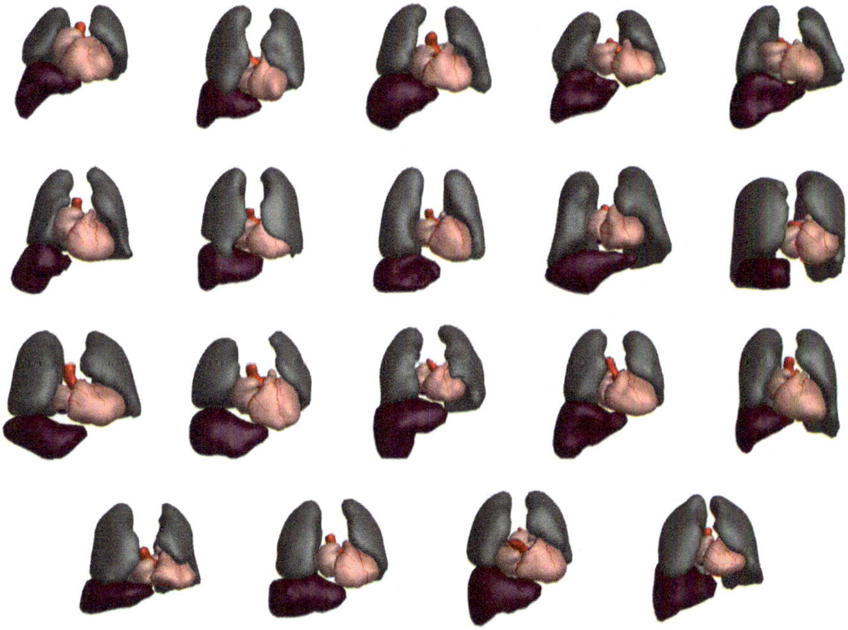

FIGURE 6.4

Examples of segmented heart (atria and ventricles), lungs, and liver from various patients. The left anterior descending (LAD) runs anteriorly over the septum, indicating the epicardial separation between the left and right ventricles. The LAD has also been segmented for all hearts. Notice the differences in heart, lungs, and liver size.

any point within or at the border of the torso, a weighted sum (integral) of the potentials at the boundaries has to be added up to Φ_∞ [31]:

$$\Phi(\vec{r}') = \Phi = \Phi_\infty(\vec{r}') - \frac{1}{4\pi}\int_{\substack{bounded\\ medium}} \Phi(r_b)\, d\omega \tag{6.1}$$

In this equation, $\Phi_\infty(\vec{r}')$ is the potential in an observation point anywhere in the infinite medium at a distance r' from the source. The potential at this observation point is affected by the potential on the boundary of the inhomogeneous element (1) and depends both on the position $\vec{r_b}$ and the potential distribution over the inhomogeneous surface, the integral of (1).

In numerical form, the integral over the surface is replaced by a summation of the potentials at a distance $\vec{r_b}$. For each observation point $(\vec{r}')$, the potential reads as:

$$\varphi(\vec{r}') = \varphi_\infty(\vec{r}') + m_r^T \varphi_{boundary} \tag{6.2}$$

In Eq. (6.2), m_r^T is a row vector in which the elements are the solid angles subtended by a small area around positions $\vec{r_b}$ at the observation point $\vec{r}'$. When the potential is calculated on the boundary, that is, $\vec{r}'$ equals $\vec{r_b}$, Eq. (6.2) can be written as a matrix multiplication [31]:

$$\varphi_{\text{boundary}} = \varphi_{\infty} + A\varphi_{\text{boundary}} = (\mathrm{I} - \mathrm{A})^{-1}\varphi_{\infty} = B\varphi_{\infty} \tag{6.3}$$

With the elements of the rows of A, a_i are the solid angles subtended by the area around all nodes on the boundary at node i.

From Eq. (6.3) follows that matrix B can be determined beforehand, due to the fact that it is determined by the boundaries of the geometries and the conductivity of these geometries. With the BEM, the transformation of Φ_{∞} to the potential in the bounded medium is reduced to a mere matrix multiplication.

6.5 Inverse problem

Solving the inverse problem in electrocardiography requires the above-described models in Fig. 6.3, which estimate the current produced by the heart from a patient-specific heart torso model (Fig. 6.4) and the potentials measured on the body surface. Solving the implied inverse problem may accordingly be viewed as a type of functional imaging, which has led to their characterization as "non-invasive electrocardiographic imaging" [32] or "myocardial activation imaging" [33].

There is a serious complication involved in the use of both models: the associated inverse problem is ill-posed, that is, small deviations in the measured ECG data may result in completely different outcomes of activation and recovery times [34]. A solution to the inverse problem can, therefore, be obtained only by regularizing the source parameters so that they express the desired properties. In the inverse solution, using the EDL as the source model, a TMP template is used to simulate ECG signals. The TMP shape is defined by the depolarization and repolarization times, as also simulated by ECGsim [35–37]. Due to the nonlinear relationship between depolarization time and repolarization time, and the shape of the TMP, the inverse problem is nonlinear; doubling the depolarization time does not result in a doubling of the simulated potentials. The advantage of the EDL, however, is its close link to electrophysiology. Using this knowledge, the required initial estimate has been used to estimate the electrical activation and recovery of the heart [12]. Understanding the physiology of the heart is thus key to improving the inverse solutions using the EDL, with which the repolarization process can also be mapped [38].

6.6 Physiology of the heart

6.6.1 Activation of the heart

The activation sequence of the human heart has been studied extensively for decades, both invasively [7,39–41] and noninvasively [42–45]. This knowledge has been used to tune the fastest-route algorithm [30] to simulate realistic cardiac activation sequences from any discrete point on the triangulated surface mesh of the atrium or ventricle [46]. The triangulated surface of the atria or ventricles is a fully interconnected graph. The vertices of the triangles are discrete points on the surface of the atria, referred to as nodes, which are connected via the edges of the triangles. The time needed for the activation wave to travel from one node to the other is proportional to the distance of the shortest path between the two nodes, when we assume that the conduction velocity is constant throughout the atrial myocardium. For the ventricles, the transmural conduction velocity is reduced due to the orientation of the myocardial fibers and the preferential conduction in the longitudinal direction of the myocardial cells [1].

Inputs to the fastest-route algorithm are the distances between adjacent nodes, stored in the adjacency matrix. The algorithm begins by searching for the closest neighbor to the focus node in the adjacency matrix. Once this node is identified, it is fixed. The distance and route to this node are recorded, and the two fixed nodes thus have a known distance. From these two nodes, the nearest neighbors are identified. The neighbor with the shortest distance to the focus is added to the list of fixed nodes, and its path and distance are stored. The distance from a fixed-node neighbor is calculated by adding the distance from the fixed node to the focus plus the fixed node → neighbor distance (Fig. 6.5).

The results from several studies have shown that this relatively simple approach to simulate the activation of the atria and ventricles matches the real cardiac activation relatively well: for normal sinus rhythm in humans [12,15], in pigs [13], as well as in ectopic activations [22,47–51]. Fastest-route algorithms enable optimization of local inhomogeneities that influence local conduction velocity, but this requires further research [52].

6.7 Recovery of the heart: ionic and molecular basis of ventricular repolarization heterogeneity

Transmural gradient of repolarization: In the hearts of most experimental animals and human cardiomyocytes, transmural variability in ion currents has been shown. Differences in repolarizing ion currents across the ventricular wall create a specific pattern of action potential duration. Typically, three cell types are recognized: epicardial, endocardial, and M cells [53]. The longer repolarization duration in endocardial cardiomyocytes compared with epicardial cells has been demonstrated in many studies [54–57].

Studies focusing on the current distribution across the free wall support the observed heterogeneity in transmural repolarization. The transient outward potassium current (I_{to}) shows transmural differences, being more prominent in ventricular epicardial than endocardial cardiomyocytes in many experimental models, including canine [58], rabbit [59], murine, and human [60–62]. In human cardiomyocytes, the transmural gradient of I_{to} is associated with a steep gradient in

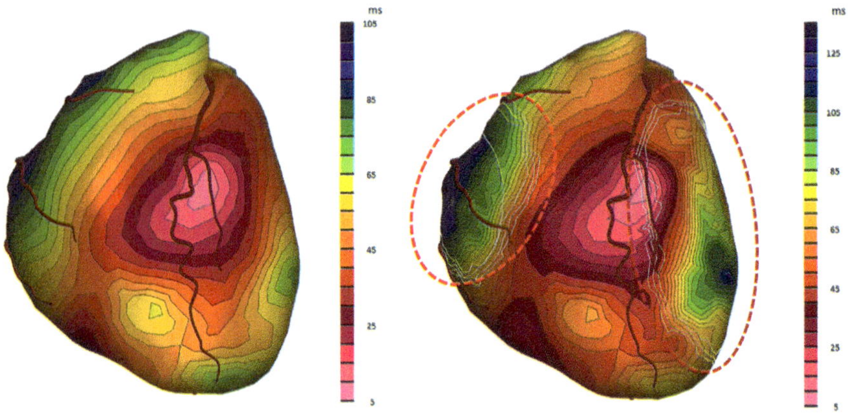

FIGURE 6.5

An example of the capability of the fastest-route algorithm. In the left panel, the activation of the ventricles for a normal His-Purkinje–initiated activation is shown. In the right panel, this activation has been manipulated by increasing the distance of the connected nodes in the two red circles by 60%. The resulting activation wave is now significantly slower in the selected areas, producing blue areas (late activation time; see color bar).

KChIP2 mRNA expression across the ventricular free wall, with KChIP2 mRNA 25 times more abundant in the epicardium than in the endocardium [63–65].

Predominance of delayed rectifier currents (I_K) in the epicardium was demonstrated for mammal and human cardiomyocytes. In the feline model, both fast and slow components (IKr, IKs) had higher densities in epicardial cardiomyocytes compared to endocardial [66]; however, in murine [60] and canine [67] cardiomyocytes, the gradient was due to the slow component (IKs).

It is generally considered that the transmural difference in repolarization duration is mainly due to the fast component of the delayed rectifier current (Ikr), whereas the slow component affects repolarization duration only when the fast component is inhibited, thereby preventing early afterdepolarizations [68].

Compared with endocardial cardiomyocytes, a greater sodium–potassium ATPase current (INaK) [69] and a larger sodium–calcium exchanger current [70] were observed in epicardial cardiomyocytes, whereas the L-type calcium current (ICa-L) was smaller in epicardial cells compared with endocardial cells [71].

A distinct type of cardiomyocyte, characterized by specific electrophysiological properties, was found in the intramural layers of the ventricular wall and was described as M cells. These cells are characterized by the highest current density of the sodium–calcium exchange mechanism [70], high Ito density, minimal IKs density, long action potential duration, and strong dependence on stimulation frequency. Distinct electrophysiological characteristics of M cells, namely prolonged action potential duration, are predominantly detected at low stimulation frequencies, whereas at physiological stimulation frequencies, the transmural gradient of repolarization is significantly reduced [72–74]. Nonetheless, the presence of M cells in vitro was confirmed in canines [75], pigs [76], felines [66], and humans [77]. Similar transmural heterogeneity of repolarization was demonstrated for the free wall of the right ventricle [63,73,78] and the interventricular septum [63,79].

It is important to emphasize that the presence of a transmural repolarization gradient observed in isolated cardiomyocytes and myocardial preparations is well established and not subject to dispute. However, evidence regarding transmural heterogeneity in repolarization duration remains inconsistent across studies conducted in in situ settings. Studies have demonstrated the absence of a transmural repolarization gradient [80–84]. A transmural repolarization gradient was observed in situ in the heart at unphysiologically low stimulation frequencies [85,86]. In studies on rabbit hearts in situ under sinus rhythm, transmural gradient of repolarization was found in the apical part of the left ventricular wall only; there were no transmural repolarization gradients in the apical part of the interventricular septum and free wall of the right ventricle, nor in the basal part of the free walls of the ventricles and the interventricular septum [87]. The reduction in differences in repolarization duration within a transmural axis could be explained by the electrotonic interactions between cardiomyocytes via gap junctions [88].

Apicobasal repolarization gradient: A repolarization sequence from the apex to the base of the ventricles was observed in many experimental models, including rabbits [89], guinea pigs [90], chickens [91], and rats [92]. The shorter repolarization duration in the apex compared to the base provides the apicobasal repolarization sequence. A corresponding distribution of repolarization durations, with longer repolarization in the ventricular base compared to the apex, has been observed in cavy [90,93–97], murine [98–100], and rabbit [89,101,102] experimental models.

Apicobasal repolarization gradient for the porcine [103], canine [104], and human [32] heart has been demonstrated. However, given the sensitivity of repolarization to methodological factors such as temperature, stimulation rate, and anesthetics, the apicobasal sequence of repolarization has been argued in other studies. Repolarization duration in the apex exceeded that in the base of ventricles in canine [105], porcine [106], and human [55] cardiomyocytes.

Differences in repolarization duration between the apical and basal areas of the ventricles were explained by ionic currents underlying repolarization. In canine cardiomyocytes from the apex, the transient output current (Ito) had higher densities as compared to basal cardiomyocytes [104]. In murine cells, the predominance of both the fast and the slow components of this current (Ito, f and Ito, s) has been established in the apex of the heart [60]. Both fast [68,107] and slow [100,104] components of delayed rectifier currents (IKs, IKr) demonstrated high density in the apex. In rabbit cardiomyocytes, predominance of rapid delayed rectifier K^+ current (Ikr) in apex and slow delayed rectifier K^+ current (IKs) in the base was found, which is in line with apicobasal gradient of repolarization in the rabbit heart in situ [89], and it corresponds to the theory that the rapid delayed rectifier potassium current (Ikr) is a primary determinant of action potential duration, whereas the slow component (IKs) provides a secondary, reserve contribution to repolarization.

Interventricular repolarization gradient: The longer action potentials were recorded in the right compared to the left ventricle in humans [108], pigs [106], and rabbits [101]. An opposite interventricular repolarization gradient had been demonstrated in canine [82,83,109,110] and murine [92,99] experimental models.

Transient outflow current (Ito) had predominance in the right compared to the left ventricle in canine [110] and murine [60,100,111] cardiomyocytes. In canine cardiomyocytes, the delayed rectifier potassium current (IK) was also demonstrated to be greater in the right than in the left ventricle [110].

Therefore, it has been evidenced that ventricular myocardium is characterized by electrophysiological heterogeneity. This repolarization heterogeneity within the ventricles is responsible for the electrocardiographic T wave. Despite a detailed description of the molecular basis for the transmural repolarization gradient, its magnitude in physiological conditions has not been fully established. The apicobasal and interventricular gradients have been studied less extensively, and the results are often controversial. The simulation models provided by the forward model ECGsim can be used to test various hypotheses about gradients in the normal healthy heart [38].

6.8 PathECG: from ECG to anatomical position

In clinical practice, the standard 12-lead ECG is widely used to diagnose patients with underlying heart disease. This interpretation is largely defined by interval measurements, that is, on the horizontal time axis (Fig. 6.6). These interval measurements seem easy for a normal ECG but can be very difficult when the ECG is abnormal. Moreover, the performance of several ECG annotation algorithms varies significantly, although all remain within the norms for ECG machines [112]. None of the ECG machines, however, supports the physician in establishing the link between the ECG and the cardiac anatomy.

The VCG was developed with this anatomical connection in mind [17], where the direction of electrical activation and recovery was reconstructed from ECG signals measured on the body surface. Several attempts were made to obtain the "true" VCG [113]. The 3D anatomical complexity and its electrical activation and recovery can never be fully captured by a single vector, not even a moving one. The clinical strength of the VCG, however, is that it provides a global view of the heart's electrical processes, which can sometimes be more informative than interpreting the combination of 12 ECG waveforms.

One of the most complex tasks in interpreting the VCG is that it is given in "body coordinates," which means the VCG interpreter needs to rotate the signals slightly to align with the heart's orientation (see Fig. 6.7A and B). This solution,

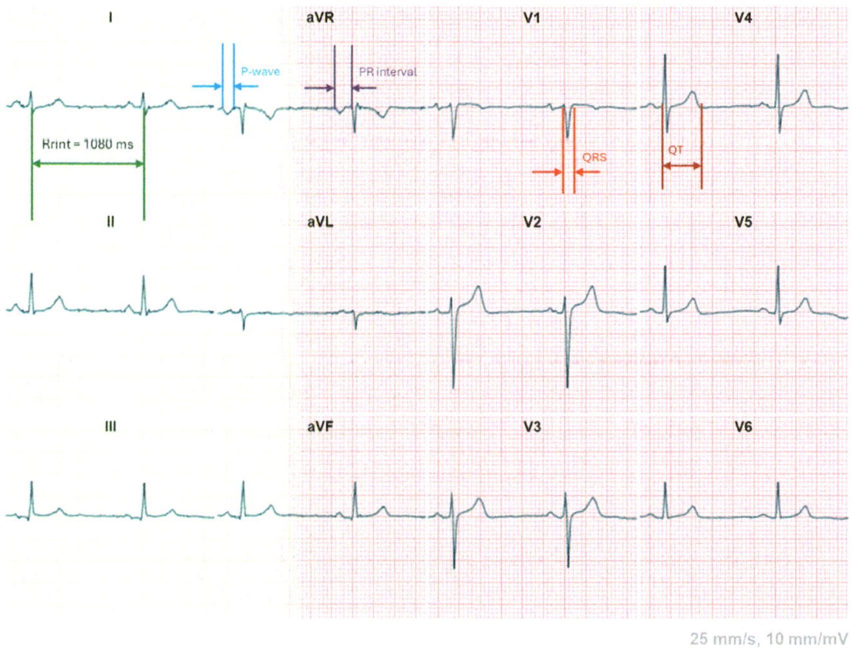

FIGURE 6.6

The standard 12-lead ECG displays the most important diagnostic measurements. From left to right: RR interval—time between two successive QRS peaks; P-wave duration—a measure of atrial activation duration; PR interval—the time from P-wave onset to QRS onset; QRS duration—the time from QRS onset to QRS end; and finally, the QT interval—from QRS onset to T-wave end. All values are traditionally measured on millimeter paper at a standard speed of 25 mm/s.

rotating the VCG to the heart axis, is a simple computer-vision rotation and supports interpretation of the QRS and T-wave directions. However, the areas with small amplitudes remain almost invisible, as in the 12-lead ECG in Figure 6.6. This is already true for the T-wave VCG. Recently, the PathECG has been proposed as an algorithm to model the movement of the VCG origin [114,115]. In short:

1) Construct the VCG from the 12-lead ECG

$$\overrightarrow{VCG}(t) = \sum_{el=1}^{9} ecg_{el}(t)\left(\frac{\overrightarrow{r_{el}} - \overrightarrow{r_{ref}}(t)}{\|\overrightarrow{r_{el}} - \overrightarrow{r_{ref}}(t)\|}\right)$$

2) Construct the PathECG from the VCG

a. Forward

$$\overrightarrow{PathECG}(t) = \overrightarrow{PathECG}(t-1) + speed\,\frac{\overrightarrow{VCG}(t)}{\|\overrightarrow{VCG}(t)\|}$$

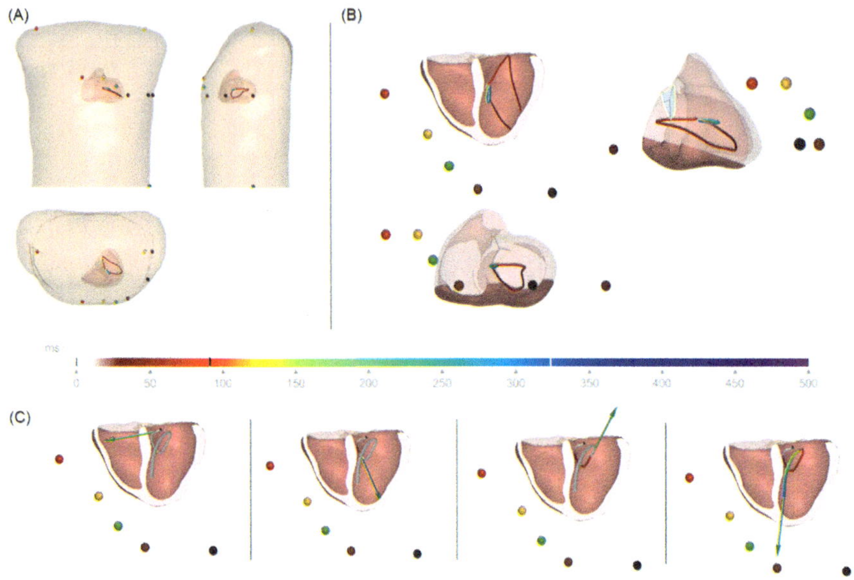

FIGURE 6.7

The movement of the heart vector. (A) The VCG over time shown in body coordinates. (B) The VCG over time shown in heart coordinates (4-chamber, right anterior obique [RAO], and LAO views). (C) The PathECG (movement of the heart vector) over time in the 4-chamber view with the heart vector at 10 ms, 30 ms, 75 ms, and end of the T wave. The color bar indicates the time from QRS onset, with white, the initial QRS; red, the QRS; and blue, the end of the T wave. The precordial ECG electrode positions are indicated (red = V1, yellow = V2, green = V3, brown = V4, black = V5, and dark brown = V6). The PathECG and VCG (B and C) can easily be related to the cardiac anatomy.

b. Backward

$$\overrightarrow{PathECG}(t) = \overrightarrow{PathECG}(t+1) - speed\,\frac{\overrightarrow{VCG}(t)}{\|\overrightarrow{VCG}(t)\|}$$

Note that the last term in each equation normalizes the length of the directional change to 1, thus assuring that the propagation speed is determined by the chosen value of speed. This procedure is repeated forward and backward until all time instants in the QRS complex have been used. The value of speed is set such that for normal activations (QRS duration < 120 ms) the PathECG stays within the heart.

Using this PathECG computation, we obtain a normal heart vector movement. The initial QRS of the PathECG in panel B shows a clear transseptal vector direction, that is, from left to right. Next, the PathECG moves to the left apex, while the terminal QRS moves to the base of the left ventricle. The T wave is directed towards the apical region, which is found in most of the normal T waves [116].

This PathECG representation thus provides a direct anatomical view of the ECG. This anatomical view is very helpful for understanding ECG waveforms, especially in 3D, in relation to the heart vector and the ECG electrodes. For instance, the detection of the combination of right bundle branch block in combination with a left anterior fascicular block is easier to detect using the PathECG [117] (Fig. 6.8).

Other applications for which the PathECG showed that it can be supportive for the diagnosis from the 12-lead ECG are:

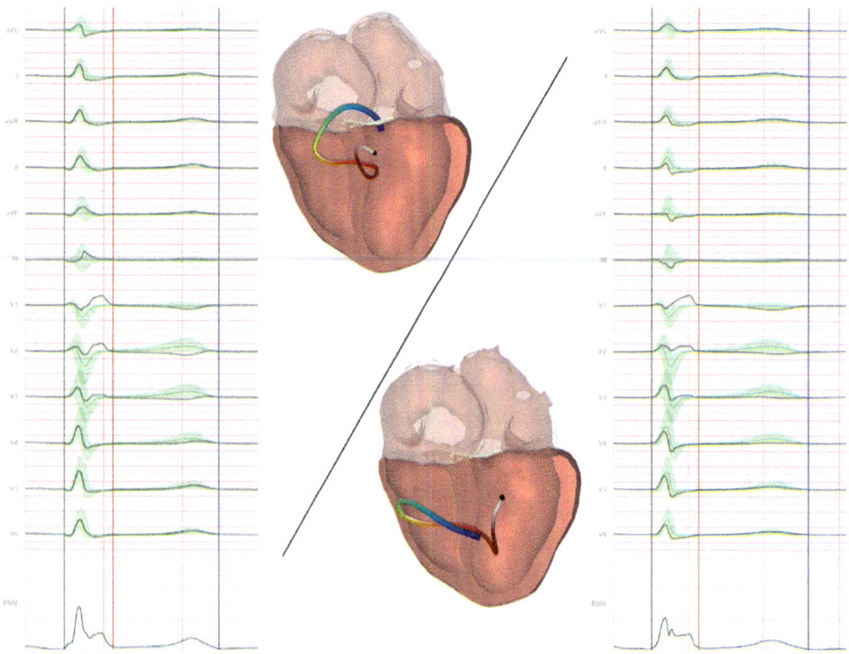

FIGURE 6.8

The PathECG from two ECG beats with similar morphology indicative of a right bundle branch block. The PathECG on the left starts near the left side of the septum with a clear transseptal activation (in white) and a late activation of the right ventricle (red–yellow PathECG) shown by the PathECG. For the ECG beat on the right, the PathECG ends also towards the right ventricles, but the initial PathECG (white) starts near the posterior septum, indicating that the left anterior bundle is also blocked for this beat.

- Brugada syndrome patients [118]
- Arrhythmogenic cardiomyopathy [119]
- Congenital ST-segment depression patients [120]
- Ischemia detection [121,122]
- Conduction disorders [123]
- Repolarization gradients in cardiac resynchronization therapy (CRT) patients [124]
- P-wave interpretation [125,126]

The PathECG shows that the standard 12-lead ECG shows more than one would expect at first glance. The PathECG also supports teaching ECG waveforms, with a better cause-and-effect relationship than memorizing standard patterns, because every patient is different. Consequently, many mistakes are made in ECG-derived diagnoses, even by cardiology experts [127]. The additional view as provided by the PathECG might reduce the number of missed diagnoses.

6.9 Conclusions

This chapter presents an integrated view of cardiac electrophysiology, combining anatomical structure, physiological function, and computational modeling approaches. It begins by describing how electrical activation governs the mechanical behavior of the heart, where depolarization initiates contraction and repolarization enables relaxation. The synchronized activity of myocardial cells is achieved through electrical coupling via gap junctions, allowing efficient propagation of excitation throughout the heart. A 3D heart model derived from computed tomography (CT) is used to illustrate the spatial organization of cardiac chambers and the conduction system.

Cardiac activation begins in the sinus node and spreads through the atria to the atrioventricular (AV) node, then conducts rapidly via the His-Purkinje system to ensure synchronized ventricular contraction. This physiological basis supports the modeling frameworks discussed in Sections 6.2–6.5. The forward problem in electrocardiography explains how electrical sources in the heart generate measurable potentials on the body surface, while source models represent cardiac electrical activity. The volume-conductor model considers how surrounding tissues influence signal propagation. Conversely, the inverse problem involves reconstructing cardiac electrical activity from measured signals, which is inherently complex and ill-posed.

Section 6.6 revisits the physiological basis of cardiac function, linking structural and electrical properties. Section 6.7 explores the ionic and molecular mechanisms underlying ventricular repolarization heterogeneity, which plays a critical role in cardiac recovery and arrhythmogenesis. Finally, Section 6.8 introduces the PathECG approach, which bridges electrocardiographic signals to their anatomical origins, enabling improved localization of cardiac activity. Together, these components provide a comprehensive framework for understanding, modeling, and interpreting cardiac electrophysiology.

References

[1] Kléber AG, Rudy Y. Basic mechanisms of cardiac impulse propagation and associated arrhythmias. Physical Review 2004;84(2):431–88.
[2] Anderson KR, Ho SY, Anderson RH. Location and vascular supply of sinus node in human heart. British Heart Journal 1979;41(1):28–32.
[3] Sanchez-Quintana D, Davies DW, Ho SY, Oslizlok P, Anderson RH. Architecture of the atrial musculature in and around the triangle of Koch: its potential relevance to atrioventricular nodal reentry. Journal of Cardiovascular Electrophysiology 1997;8(12):1396–407.
[4] Spach MS, Lieberman M, Scott JG, Barr RC, Johnson EA, Kootsey JM. Excitation sequences of the atrial septum and the AV node in isolated hearts of the dog and rabbit. Circulation Research 1971;29:156–72.
[5] Anderson RH, Ho SY. The architecture of the sinus node, the atrioventricular conduction axis, and the internodal atrial myocardium. Journal of Cardiovascular Electrophysiology 1998;9(11):1233–48.
[6] Demoulin JC, Kulbertus HE. Histopathological examination of concept of left hemiblock. British Heart Journal 1972;34(8):807–14.
[7] Durrer D, van Dam RT, Freud GE, Janse MJ, Meijler FL, Arzbaecher RC. Total excitation of the isolated human heart. Circulation 1970;41:899–912.
[8] Conrath CE, Wilders R, Coronel R, de Bakker JMT, Taggart P, de Groot JR, et al. Intercellular coupling through gap junctions masks M cells in the human heart. Cardiovascular Research 2004;62(2):407–14.
[9] Narayan SM, John RM. Advanced electroanatomic mapping: current and emerging approaches. Current Treatment Options in Cardiovascular Medicine 2024;26(4):69–91.
[10] Vigmond EJ, Massé S, Roney CH, Bayer JD, Nanthakumar K. The accuracy of cardiac surface conduction velocity measurements. JACC: Clinical Electrophysiology 2025;11(4):694–705.
[11] Oostendorp TF, Cluitmans M, Coronel R, Dubois R. Adriaan van Oosterom, PhD. Heart Rhythm 2019;16(10):e299.
[12] van Dam PM, Oostendorp TF, Linnenbank AC, van Oosterom A. Non-invasive imaging of cardiac activation and recovery. Annals of Biomedical Engineering 2009;37(9):1739–56.
[13] Oosterhoff P, Meijborg VMF, van Dam PM, van Dessel PFHM, Belterman CNW, Streekstra GJ, et al. Experimental validation of noninvasive epicardial and endocardial activation imaging. Circulation: Arrhythmia Electrophysiology 2016;9(8):e004104.
[14] Roudijk RW, Boonstra MJ, Brummel R, Kassenberg W, Blom LJ, Oostendorp TF, et al. Comparing non-invasive inverse electrocardiography with invasive endocardial and epicardial electroanatomical mapping during sinus rhythm. Frontiers in Physiology 2021;12(1625).
[15] Boonstra MJ, Roudijk RW, Brummel R, Kassenberg W, Blom LJ, Oostendorp TF, et al. Modeling the His-Purkinje effect in non-invasive estimation of endocardial and epicardial ventricular activation. Annals of Biomedical Engineering 2022;50(3):343–59.
[16] Wilson FN, Macleod AG, Barker PS. The distribution of action currents produced by the heart muscle and other excitable tissues immersed in conducting media. Journal of General Physiology 1933;16:423–56.
[17] Wilson FN, Johnston FD. The vectorcardiogram. American Heart Journal 1938;16(1):14–28.

[18] van Oosterom A. A comparison of electrocardiographic imaging based on two source types. Europace 2014;16(Suppl 4):iv120–8.
[19] van Oosterom A. The dominant T wave and its significance. Journal of Cardiovascular Electrophysiology 2003;14(10 Suppl):S180–7.
[20.] van Oosterom A. The dominant T wave. Journal of Electrocardiology 2004;37(Suppl):193–7.
[21] van Oosterom A. Reflections on T waves. In: Hiraoka M, Ogawa S, Kodama I, Kasnuki H, Katoh T, editors. Advances in electrocardiology. New Jersey: World Scientific; 2005. p. 807–15.
[22] van Dam PM, Boyle NG, Laks MM, Tung R. Localization of premature ventricular contractions from the papillary muscles using the standard 12-lead electrocardiogram: a feasibility study using a novel cardiac isochrone positioning system. Europace 2016;18(Suppl. 4):iv16–22.
[23] van Oosterom A. Genesis of the T wave as based on an equivalent surface source model. Journal of Electrocardiography 2001;34(Suppl. 2001):217–27.
[24] van Oosterom A, Jacquemet V. Genesis of the P wave: atrial signals as generated by the equivalent double layer source model. Europace 2005;7(s2):S21–9.
[25] Geselowitz DB. On the Theory of the electrocardiogram. Proceedings of the IEEE 1989;77(6):857–76.
[26] Geselowitz DB. Description of cardiac sources in anisotropic cardiac muscle. Application Bidomain Model Journal of Electrocardiology 1992;25(Suppl):65–7.
[27] Janssen AM, Potyagaylo D, Dossel O, Oostendorp TF. Assessment of the equivalent dipole layer source model in the reconstruction of cardiac activation times on the basis of BSPMs produced by an anisotropic model of the heart. Medical and Biological Engineering and Computing 2018;56(6):1013–25.
[28] van Oosterom A, Jacquemet V. A parameterized description of transmembrane potentials used in forward and inverse procedures. International Conference of Electrocardiology. Gdansk; Poland: Folia Cardiologica; 2005.
[29] van Oosterom A, van Dam PM. editors. The intra-myocardial distance function used in inverse computations of the timing of depolarization and repolarization. Computers Cardiology 2005:25–8. Sept.
[30] van Dam PM, Oostendorp TF, van Oosterom A. Application of the fastest route algorithm in the interactive simulation of the effect of local ischemia on the ECG. Medical and Biological Engineering and Computing 2009;47(1):11–20.
[31] van Oosterom A, Oostendorp TF. Electric volume conduction. Nijmegen: not published, 2002.
[32] Ramanathan C, Jia P, Ghanem R, Ryu K, Rudy Y. Activation and repolarization of the normal human heart under complete physiological conditions. PNAS 2006;103(16):6309–14.
[33] Huiskamp GJM, Greensite F. A new method for myocardial activation imaging. IEEE Transactions on Biomedical Engineering 1997;44(6):433–46.
[34] Gulrajani RM. The inverse problem in electrocardiography. Bioelectricity and biomagnetism. New York: John Wiley & Sons; 1998. p. 381–431.
[35] van Dam PM, Oostendorp TF, van Oosterom A. ECGSIM: interactive simulation of the ECG for teaching and research purposes. Computing in Cardiology 2010;37:841–4.
[36] van Oosterom A, Oostendorp TF, van Dam PM. Potential applications of the new ECGSIM. Journal of Electrocardiology 2011;44(5):577–83.

[37] van Dam PM, van Dam EM, van Oosterom A, Oostendorp TF. Interactive simulation of multiple beats: a new feature of ECGSIM. Computing in cardiology. Boston: IEEE Computer Society Press; 2014.

[38] Kloosterman M, van der Schaaf I, Boonstra MJ, Oostendorp TF, Meijborg VMF, Coronel R, et al. Genesis of the T-wave through various modes of ventricular recovery patterns using the equivalent dipole layer model. Computers in Biology and Medicine 2025;189:110016.

[39.] De Ponti R, Ho SY, Salerno-Uriarte JA, Tritto M, Spadacini G. Electroanatomic analysis of sinus impulse propagation in normal human atria. Journal of Cardiovascular Electrophysiology 2002;13(1):1–10.

[40] Wyndham CRM, Smith T, Saxema A, Engleman S, Levitsky RM, Rosen KM. Epicardial activation of the intact human heart without conduction defect. Circulation 1979;59(1):161–8.

[41] Hill AJ, Laske TG, Coles JJA, Sigg DC, Skadsberg ND, Vincent SA, et al. In vitro studies of human hearts. Annals of Thoracic Surgery 2005;79(1):168–77.

[42] Huiskamp GJH, van Oosterom A. The depolarization sequence of the human heart surface computed from measured body surface potentials. IEEE Transactions on Biomedical Engineering 1988;35(12):1047–58.

[43] Huiskamp GJM, Oostendorp TF, Pijls NHJ, van Oosterom A. Invasive confirmation of the human ventricular activation sequence as computed from body surface potentials. Los Alamitos: IEEE Computer Society Press; 1993. Computers in Cardiology92 editors. Invasive confirmation of the human ventricular activation sequence as computed from body surface potentials.

[44] Ghanem RN, Jia P, Ramanathan C, Ryu K, Markowitz A, Rudy Y. Noninvasive electrocardiographic imaging (ECGI): comparison to intraoperative mapping in patients. Heart Rhythm 2005;2(4):339–54.

[45] Fischer G, Hanser F, Pfeifer B, Seger M, Hintermuller C, Modre R, et al. A signal processing pipeline for noninvasive imaging of ventricular preexcitation. Methods of Information in Medicine 2005;44:588. 15.

[46] Potyagaylo D, Dam DO. PV. Influence of modeling errors on the initial estimate for nonlinear myocardial activation times imaging calculated with fastest route algorithm. IEEE Transactions on Biomedical Engineering 2016;(99): 1.

[47] Misra S, van Dam PM, Chrispin J, Assis F, Keramati A, Kolandaivelu A, et al. Initial validation of a novel ECGI system for localization of premature ventricular contractions and ventricular tachycardia in structurally normal and abnormal hearts. Journal of Electrocardiology 2018;51(5):801–8.

[48] Roudijk RW, Boonstra MJ, Van Dam PM, Van Der Heijden JF, Asselbergs FW, Loh KP. In-vivo validation of non-invasive electrocardiographic imaging based on the equivalent double layer source model using epicardial and endocardial electroanatomical mapping. European Heart Journal 2020;41(Supplement_2).

[49] Melgaard J, van Dam PM, Sommer A, Fruelund P, Nielsen JC, Riahi S, et al. Non-invasive estimation of QLV from the standard 12-lead ECG in patients with left bundle branch block. Frontiers in Physiology 2022;13.

[50] Fruelund PZ, Van Dam PM, Melgaard J, Sommer A, Lundbye-Christensen S, Søgaard P, et al. Novel non-invasive ECG imaging method based on the 12-lead ECG for reconstruction of ventricular activation: a proof-of-concept study. Frontiers in Cardiovascular Medicine 2023;10.

[51] Sedova KA, van Dam PM, Blahova M, Necasova L, Kautzner J. Localization of the ventricular pacing site from BSPM and standard 12-lead ECG: a comparison study. Science Reports 2023;13(1):9618.
[52] Boonstra MJ, Oostendorp TF, Roudijk RW, Kloosterman M, Asselbergs FW, Loh P, et al. Incorporating structural abnormalities in equivalent dipole layer based ECG simulations. Frontiers in Physiology 2022;13.
[53] Antzelevitch C, Shimizu W, Yan GX, Sicouri S, Weissenburger J, Nesterenko VV, et al. The M cell: its contribution to the ECG and to normal and abnormal electrical function of the heart. Journal of Cardiovascular Electrophysiology 1999;10(8):1124–52.
[54] Abildskov JA. The sequence of normal recovery of excitability in the dog heart. Circulation 1975;52(3):442–6.
[55] Franz MR, Bargheer K, Rafflenbeul W, Haverich A, Lichtlen PR. Monophasic action potential mapping in human subjects with normal electrocardiograms: direct evidence for the genesis of the T wave. Circulation 1987;75(2):379–86.
[56] Higuchi T, Nakaya Y. T wave polarity related to the repolarization process of epicardial and endocardial ventricular surfaces. American Heart Journal 1984;108(2):290–5.
[57] van DR, Durrer D. Experimental study on the intramural distribution of the excitability cycle and on the form of the epicardial T wave in the dog heart in situ. American Heart Journal 1961;61:537–42.
[58] Furukawa T, Kimura S, Furukawa N, Bassett AL, Myerburg RJ. Role of cardiac ATP-regulated potassium channels in differential responses of endocardial and epicardial cells to ischemia. Circulation Research 1991;68(6):1693–702.
[59] Fedida D, Giles WR. Regional variations in action potentials and transient outward current in myocytes isolated from rabbit left ventricle. Journal of Physiology 1991;442:191–209.
[60] Brunet S, Aimond F, Li H, Guo W, Eldstrom J, Fedida D, et al. Heterogeneous expression of repolarizing, voltage-gated K+ currents in adult mouse ventricles. Journal of Physiology 2004;559(Pt 1):103–20.
[61] Guo W, Li H, Aimond F, Johns DC, Rhodes KJ, Trimmer JS, et al. Role of heteromultimers in the generation of myocardial transient outward K+ currents. Circulation Research 2002;90(5):586–93.
[62] Kondo RP, Dederko DA, Teutsch C, Chrast J, Catalucci D, Chien KR, et al. Comparison of contraction and calcium handling between right and left ventricular myocytes from adult mouse heart: a role for repolarization waveform. Journal of Physiology 2006;571(Pt 1):131–46.
[63] Rosati B, Grau F, Rodriguez S, Li H, Nerbonne JM, McKinnon D. Concordant expression of KChIP2 mRNA, protein and transient outward current throughout the canine ventricle. Journal of Physiology 2003;548(Pt 3):815–22.
[64] Rosati B, Pan Z, Lypen S, Wang HS, Cohen I, Dixon JE, et al. Regulation of KChIP2 potassium channel beta subunit gene expression underlies the gradient of transient outward current in canine and human ventricle. Journal of Physiology 2001;533(Pt 1):119–25.
[65] Zicha S, Xiao L, Stafford S, Cha TJ, Han W, Varro A, et al. Transmural expression of transient outward potassium current subunits in normal and failing canine and human hearts. Journal of Physiology 2004;561(Pt 3):735–48.

[66] Aiba T, Shimizu W, Inagaki M, Hidaka I, Tatewaki T, Sunagawa K. Transmural heterogeneity of the action potential configuration in the feline left ventricle. Circulation Journal 2003;67(5):449–54.
[67] Liu DW, Antzelevitch C. Characteristics of the delayed rectifier current (IKr and IKs) in canine ventricular epicardial, midmyocardial, and endocardial myocytes. A weaker IKs contributes to the longer action potential of the M cell. Circulation Research 1995;76(3):351–65.
[68] Brahmajothi MV, Morales MJ, Reimer KA, Strauss HC. Regional localization of ERG, the channel protein responsible for the rapid component of the delayed rectifier, K+ current in the ferret heart. Circulation Research 1997;81(1):128–35.
[69] Arita M, Imanishi S, Aomine M, Kiyosue T. Na,K-ATPase activity and repolarization of ventricular action potentials in simian hearts. Japan Heart Journal 1986;27(Suppl. 1):145–52.
[70] Zygmunt AC, Goodrow RJ, Antzelevitch CI. NaCa) contributes to electrical heterogeneity within the canine ventricle. American Journal of Physiology-Heart and Circulatory Physiology 2000;278(5):H1671–8.
[71] Wang HS, Cohen IS. Calcium channel heterogeneity in canine left ventricular myocytes. Journal of Physiology 2003;547(Pt 3):825–33.
[72] Liu DW, Gintant GA, Antzelevitch C. Ionic bases for electrophysiological distinctions among epicardial, midmyocardial, and endocardial myocytes from the free wall of the canine left ventricle. Circulation Research 1993;72(3):671–87.
[73] Sicouri S, Antzelevitch C. A subpopulation of cells with unique electrophysiological properties in the deep subepicardium of the canine ventricle. The M cell. Circulation Research 1991;68(6):1729–41.
[74] Sicouri S, Antzelevitch C. Electrophysiologic characteristics of M cells in the canine left ventricular free wall. Journal of Cardiovascular Electrophysiology 1995;6(8):591–603.
[75] Sicouri S, Quist M, Antzelevitch C. Evidence for the presence of M cells in the guinea pig ventricle. Journal of Cardiovascular Electrophysiology 1996;7(6):503–11.
[76] Stankovicova T, Szilard M, De Scheerder I, Sipido KR. M cells and transmural heterogeneity of action potential configuration in myocytes from the left ventricular wall of the pig heart. Cardiovascular Research 2000;45(4):952–60.
[77] Drouin E, Lande G, Charpentier F. Amiodarone reduces transmural heterogeneity of repolarization in the human heart. Journal of the American College of Cardiology 1998;32(4):1063–7.
[78] Li D, Li CY, Yong AC, Kilpatrick D. Source of electrocardiographic ST changes in subendocardial ischemia. Circulation Research 1998;82(9):957–70.
[79] Ramakers C, Stengl M, Spatjens RL, Moorman AF, Vos MA. Molecular and electrical characterization of the canine cardiac ventricular septum. Journal of Molecular Cell Cardiology 2005;38(1):153–61.
[80] Anyukhovsky EP, Sosunov EA, Feinmark SJ, Rosen MR. Effects of quinidine on repolarization in canine epicardium, midmyocardium, and endocardium: II. In vivo study. Circulation 1997;96(11):4019–26.
[81] Anyukhovsky EP, Sosunov EA, Rosen MR. Regional differences in electrophysiological properties of epicardium, midmyocardium, and endocardium. In vitro and in vivo correlations. Circulation 1996;94(8):1981–8.
[82] Coronel R, Opthof T, Plotnikov AN, Wilms-Schopman FJ, Shlapakova IN, Danilo P, Jr., et al. Long-term cardiac memory in canine heart is associated with the evolution of a transmural repolarization gradient. Cardiovascular Research 2007;74(3):416–25.

[83] Janse MJ, Sosunov EA, Coronel R, Opthof T, Anyukhovsky EP, de Bakker JM, et al. Repolarization gradients in the canine left ventricle before and after induction of short-term cardiac memory. Circulation 2005;112(12):1711–8.
[84] Taggart P, Sutton PM, Opthof T, Coronel R, Trimlett R, Pugsley W, et al. Transmural repolarisation in the left ventricle in humans during normoxia and ischaemia. Cardiovascular Research 2001;50(3):454–62.
[85] el-Sherif N, Caref EB, Yin H, Restivo M. The electrophysiological mechanism of ventricular arrhythmias in the long QT syndrome. Tridimensional mapping of activation and recovery patterns. Circulation Research 1996;79(3):474–92.
[86] Taggart P, Sutton P, Opthof T, Coronel R, Kallis P. Electrotonic cancellation of transmural electrical gradients in the left ventricle in man. Progress in Biophysics and Molecular Biology 2003;82(1-3):243–54.
[87] Arteyeva NV, Goshka SL, Sedova KA, Bernikova OG, Azarov JE. What does the T(peak)-T(end) interval reflect? An experimental and model study. Journal of Electrocardiology 2013;46(4):e1–8. 296.
[88] Conrath CE, Wilders R, Coronel R, de Bakker JM, Taggart P, de Groot JR, et al. Intercellular coupling through gap junctions masks M cells in the human heart. Cardiovascular Research 2004;62(2):407–14.
[89] Sedova KA, Goshka SL, Vityazev VA, Shmakov DN, Azarov JE. Load-induced changes in ventricular repolarization: evidence of autonomic modulation. Canadian Journal of Physiology and Pharmacology 2011;89(12):935–44.
[90] Kanai A, Salama G. Optical mapping reveals that repolarization spreads anisotropically and is guided by fiber orientation in guinea pig hearts. Circulation Research 1995;77(4):784–802.
[91] Kharin SN. Depolarisation and repolarisation sequences of ventricular epicardium in chickens (Gallus gallus domesticus). Comparative Biochemistry and Physiology Part A: Molecular and Integrative Physiology 2004;137(1):237–44.
[92] Watanabe T, Delbridge LM, Bustamante JO, McDonald TF. Heterogeneity of the action potential in isolated rat ventricular myocytes and tissue. Circulation Research 1983;52(3):280–90.
[93] Choi BR, Liu T, Salama G. The distribution of refractory periods influences the dynamics of ventricular fibrillation. Circulation Research 2001;88(5):E49–58.
[94] Choi BR, Salama G. Simultaneous maps of optical action potentials and calcium transients in guinea-pig hearts: mechanisms underlying concordant alternans. Journal of Physiology 2000;529(Pt 1):171–88. (Pt 1).
[95] Efimov IR, Ermentrout B, Huang DT, Salama G. Activation and repolarization patterns are governed by different structural characteristics of ventricular myocardium: experimental study with voltage-sensitive dyes and numerical simulations. Journal of Cardiovascular Electrophysiology 1996;7(6):512–30.
[96] Poelzing S, Veeraraghavan R. Heterogeneous ventricular chamber response to hypokalemia and inward rectifier potassium channel blockade underlies bifurcated T wave in guinea pig. American Journal of Physiology-Heart and Circulatory Physiology 2007;292(6):H3043–51.
[97] Watanabe T, Rautaharju PM, McDonald TF. Ventricular action potentials, ventricular extracellular potentials, and the ECG of guinea pig. Circulation Research 1985;57(3):362–73.
[98] Baker LC, London B, Choi BR, Koren G, Salama G. Enhanced dispersion of repolarization and refractoriness in transgenic mouse hearts promotes reentrant ventricular tachycardia. Circulation Research 2000;86(4):396–407.

[99] Donohoe P, Hendry BM, Walgama OV, Bertaso F, Hopster DJ, Shattock MJ, et al. An altered repolarizing potassium current in rat cardiac myocytes after subtotal nephrectomy. Journal of the American Society of Nephrology JASN 2000;11(9):1589–99.

[100] Guo W, Xu H, London B, Nerbonne JM. Molecular basis of transient outward K+ current diversity in mouse ventricular myocytes. Journal of Physiology 1999;521(Pt 3):587–99. (Pt 3).

[101] Azarov JE, Shmakov DN, Vityazev VA, Roshchevskaya IM, Roshchevsky MP. Activation and repolarization patterns in the ventricular epicardium under sinus rhythm in frog and rabbit hearts. Comparative Biochemistry and Physiology Part A: Molecular and Integrative Physiology 2007;146(3):310–6.

[102] Sedova KA, Vityazev VA, Azarov JE, Shmakov DN. Repolarization of epicardial ventricular surface of rabbit heart in acute stenosis of the aortic arch. Bulletin of Experimental Biology and Medicine 2008;146(2):180–1.

[103] Dean JW, Lab MJ. Regional changes in ventricular excitability during load manipulation of the in situ pig heart. Journal of Physiology 1990;429:387–400.

[104] Szentadrassy N, Banyasz T, Biro T, Szabo G, Toth BI, Magyar J, et al. Apico-basal inhomogeneity in distribution of ion channels in canine and human ventricular myocardium. Cardiovascular Research 2005;65(4):851–60.

[105] Toyoshima H, Lux RL, Wyatt RF, Burgess M, Abildskov JA. Sequences of early and late phases of repolarization on dog ventricular epicardium. Journal of Electrocardiology 1981;14(2):143–52.

[106] Newton JC, Johnson PL, Justice RK, Smith WM, Ideker RE. Estimated global epicardial distribution of activation rate and conduction block during porcine ventricular fibrillation. Journal of Cardiovascular Electrophysiology 2002;13(10):1035–41.

[107] Cheng J, Kamiya K, Liu W, Tsuji Y, Toyama J, Kodama I. Heterogeneous distribution of the two components of delayed rectifier K+ current: a potential mechanism of the proarrhythmic effects of methanesulfonanilideclass III agents. Cardiovascular Research 1999;43(1):135–47.

[108] Chen PS, Moser KM, Dembitsky WP, Auger WR, Daily PO, Calisi CM, et al. Epicardial activation and repolarization patterns in patients with right ventricular hypertrophy. Circulation 1991;83(1):104–18.

[109] Leerssen HM, Vos MA, Houben R, den Dulk K, Wellens HJ. High uniformity of left and right ventricular repolarization dynamics induced by an abrupt decrease in pacing cycle length in a dog is not affected by left ventricular ischemia. Journal of Cardiovascular Electrophysiology 2000;11(4):421–9.

[110] Volders PG, Sipido KR, Carmeliet E, Spatjens RL, Wellens HJ, Vos MA. Repolarizing K+ currents ITO1 and IKs are larger in right than left canine ventricular midmyocardium. Circulation 1999;99(2):206–10.

[111] Xu H, Guo W, Nerbonne JM. Four kinetically distinct depolarization-activated K+ currents in adult mouse ventricular myocytes. Journal of General Physiology 1999;113(5):661–78.

[112] De Bie J, Martignani C, Massaro G, Diemberger I. Performance of seven ECG interpretation programs in identifying arrhythmia and acute cardiovascular syndrome. Journal of Electrocardiology 2020;58:143–9.

[113] Burger HC, van Brummelen AGW, vanHerpen G. Comprimise in vectorcardiography ii. alterations of coefficients as a means of adapting one lead system to another. American Heart Journal 1962;64:666–78.

[114] Boonstra MJ, Brooks DH, Loh P, van Dam PM. CineECG: a novel method to image the average activation sequence in the heart from the 12-lead ECG. Computers in Biology and Medicine 2021:105128.

[115] Boonstra MJ, Hilderink BN, Locati ET, Asselbergs FW, Loh P, van Dam PM. Novel CineECG enables anatomical 3D localization and classification of bundle branch blocks. EP Europace 2021;23(Suppl. 1):i80–7.

[116] van Dam PM, Boonstra M, Locati ET, Loh P. The relation of 12 lead ECG to the cardiac anatomy: the normal CineECG. Journal of Electrocardiology 2021;69:67–74.

[117] van der Schaaf I, Kloosterman M, Boonstra MJ, van Dam PM, Gorgels APM. CineECG illustrating the ventricular activation sequence in progressive AV-junctional conduction block. Journal of Electrocardiology 2023;78:1–4.

[118] Dam PMV, Locati ET, Ciconte G, Borrelli V, Heilbron F, et al. Novel CineECG derived from standard 12-lead ECG enables right ventricle outflow tract localization of electrical substrate in patients with Brugada syndrome. Circulation: Arrhythmia Electrophysiology 2020;13(9):e008524.

[119] van der Schaaf I, Kloosterman M, Gorgels APM, te Riele ASJM, van Dam PM, Loh P. CineECG detects abnormal electrical activity in the 12-lead ECG of preclinical plakophilin-2 variant carriers. Heart Rhythm O2 2025;6(11):1786–95.

[120] Frosted R, Paludan-Muller C, Vad OB, Olesen MS, Bundgaard H, van Dam P, et al. CineECG analysis provides new insights into familial ST-segment depression syndrome. Europace 2023.

[121] Gorgels APM, van der Schaaf I, Kloosterman M, van Dam PM. The CineECG in ischemia localization in ST-elevation (equivalent) acute coronary syndromes. Journal of Electrocardiology 2023;81:258–64.

[122] van der Schaaf I, Kloosterman M, Gorgels APM, Loh P, van Dam PM. CineECG for visualization of changes in ventricular electrical activity during ischemia. Journal of Electrocardiology 2024.

[123] Mortada MJ, Sbrollini A, Marcantoni I, Iammarino E, Burattini L, Dam PV. Quantifying CineECG output for enhancing electrocardiography signals classification. IEEE Open Journal of Engineering in Medicine and Biology 2025;6:488–98.

[124] Sedova K, Azarov JE, Van Dam PM, Necasova L, Kukla J, Sramko M, et al. CineECG repolarization gradients predict acute hemodynamic response in CRT patients. Journal of Cardiovascular Electrophysiology 2024.

[125] Baturova M, Van Dam P, Johnson LSB, Smith G, Platonov P. Usefulness of CineECG method in identification of P-wave morphology variants in an elderly epidemiologic cohort. European Heart Journal 2023;44(Suppl. 2).

[126] Locati ET, Pappone C, Heilbron F, van Dam PM. CineECG provides a novel anatomical view on the normal atrial P-wave. European Heart Journal - Digital Health 2022;3(2):169–80.

[127] Cook DA, Oh SY, Pusic MV. Accuracy of physicians' electrocardiogram interpretations: a systematic review and meta-analysis. JAMA International Medicine 2020;180(11):1461.

CHAPTER

Breath biopsy for heart failure monitoring

7

Paweł Kacprzak[1,2], Michał Mączewski[3], Anna Paleczek[4], Dominik Grochala[4], Stanisław Karcz[4], Mateusz Kocoń[4] and Artur Rydosz[1,4,5]

[1]*Advanced Medical Equipment, Warsaw, Poland,*
[2]*Helgelandssykehuset HF, Mo I Rana, Norway,*
[3]*Department of Clinical Physiology, Centre of Postgraduate Medical Education, Warsaw, Poland,*
[4]*Biomarkers Analysis LAB, Institute of Electronics, AGH University of Kraków, Kraków, Poland,*
[5]*Functional and Virtual Medical 3D Imaging Laboratory, Department of Diagnostic Imaging, University Hospital in Kraków, Kraków, Poland*

7.1 Introduction

Breath biopsy, commonly known as exhaled breath analysis (EBA), involves analyzing the compounds in exhaled breath. Therefore EBA can be applied for human and animal disease screening, monitoring, and medical decision-making processes. Beyond the predominant constituents of exhaled breath—nitrogen, oxygen, and carbon dioxide—clinically relevant information is conveyed through volatile biomarkers, including ketones, aldehydes, and others. During breathing, some of the compounds present in the body are transported by blood to the lungs and then exhaled, while others are inhaled, dissolved in the blood, and transported throughout the body. In general, we can distinguish between endogenous and exogenous biomarkers; however, it is essential to keep in mind that inhaled compounds may react in the human body in both positive (e.g., drugs) and negative (e.g., carcinogenic compounds such as toluene and benzene) ways, and thus the exhaled composition may change. Therefore the exhaled composition of compounds contains metabolic information from the whole body, paving the way for using EBA for the diagnostic purposes of many organs, not just the lungs. However, the exhaled air is not homogeneous. In fact, the first part of the breath is called "dead volume" and does not correlate with blood biomarkers. Therefore the key challenge lies in collecting the final portion of exhaled breath, known as alveolar air. In some cases, mixed exhaled air is collected and analyzed instead, as it is more convenient for both patients and researchers. Regardless of the EBA methodology, ambient air should also be interpreted as a background reference.

In the case of exhaled breath biomarkers, the primary concerns relate to the biological, physiological, or pathological pathway that enables the association between the biomarker's presence and/or concentration and the disease or condition

Functional and Virtual Medical 3D Imaging in Cardiovascular Medicine. DOI: https://doi.org/10.1016/B978-0-443-44570-5.00004-5

of a specific disorder. Discovering such a linkage between a biomarker and its correlation to the health condition paves the way for noninvasive diagnosis. However, it must be understood from the outset that shared biochemical pathways in the body may lead to a typical biological or clinical phenotype among diseases, as well as the appearance of the same exhaled compound in various diseases. Therefore, in most cases, it is not possible to establish a simple linkage between a single exhaled biomarker and a single disease, similar to clinical examination, where several factors are considered before a diagnosis is made.

7.1.1 Breath biopsy—a new term

Breath biopsy is a new term introduced by the Owlstone Medical Company (OMC) [1]. It refers to the OMC products and services that provide a reliable platform for collecting and analyzing breath samples. Although the term is primarily associated with OMC products, it may be adapted to encompass all EBA in the future, based on biopsies of other organs from a medical perspective. In general, the "breath biopsy" term may be used for all tools used for noninvasive screening of various diseases, stratifying patients by phenotype for precision medicine, monitoring response to treatment of specific diseases such as cardiovascular disease, and for analysis of health conditions of subjects under exposure to hazardous compounds, etc. However, before a new term is implemented in the daily clinical routine, as well as a new biomarker or a set of biomarkers used for diagnosis, monitoring, and treatment, a few criteria must be fulfilled. In 2007, Morrow and de Lemos [2] proposed three criteria that can be explained as accuracy, novelty, and applicability in clinical routines. The accuracy of the measurement method refers to its repeatability, achieved at a reasonable cost and with short turnaround times. Novelty should provide a process that extends beyond what is already available from a careful clinical assessment—for example, exhaled acetone measurement, which is inexpensive, easy to conduct, well-correlated with blood biomarkers such as N-terminal pro-B-type natriuretic peptide (NT-proBNP), and the New York Heart Association (NYHA) classification [2]. Finally, the applicability of the daily clinical routine should always be kept in mind when researchers start working on new methods. From this perspective, engineers and physicians should work hand in hand. Otherwise, the devices and methods engineers develop will never be used by physicians, and vice versa. The methods that physicians want cannot see the daylight if engineers are not informed about the physicians' needs.

7.1.2 Breath sampling methods

There are several variations of the exhaled breath sampling (EBS) procedures; however, from the patient's perspective, the collection or sampling should be conducted comfortably; for example, the subject should be seated, remain at rest during the procedure, and breathe normally. Currently, there is no consensus on the requirement for fasting, especially regarding the duration of fasting (i.e., how many hours subjects should abstain from eating and drinking). However, avoiding consuming alcohol and using tobacco in any form (including the currently popular vape and electronic cigarettes) is commonly accepted and used in all EBA methods. In 2014, Herbig and Beauchamp [3] pointed out that a universal standard procedure for EBS is not possible due to several reasons, including the rich chemical diversity of exhaled breath that covers an extensive range of volatile organic compounds (VOCs) at highly varied concentrations; the vast disparity in the analytical tools employed; diverse study goals; and the presence of unidentified confounders [3]. To date, two major approaches are used for EBS, that is, mixed and alveolar. In the first one, the total breath, including dead space air, is collected. In contrast, the former approach collects only the alveolar part of exhalation, as illustrated in Fig. 7.1. From a practical point of view, collecting and analyzing mixed samples is easier for both patients and medical staff. Spontaneous breathing without additional equipment and preparation seems to be the only entirely acceptable maneuver for EBA. However, due to the dead volume, analyzing biomarkers at low concentrations is very challenging. On the other hand, the alveolar samples are initially prepared for

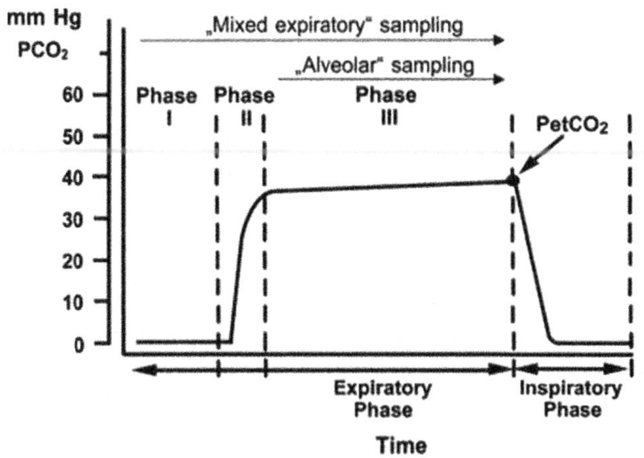

FIGURE 7.1 Schematic drawing of a normal capnogram and typical modes of sampling.

Phase I is the first expiratory stage, and the gas sampled during this phase represents anatomical dead space, typically not containing CO_2 or endogenous VOCs. The appearance of CO_2 and a steep upstroke in CO_2 tension in the normal capnogram characterizes phase II. Gas sampled during this phase typically includes a mixture of alveolar and dead space air. Phase III reflects a minimal change in CO_2 concentration resulting from alveolar emptying. This phase is referred to as the alveolar plateau or expiratory plateau. $PetCO_2$ is the terminal portion of exhaled carbon dioxide, which reveals the actual end-tidal concentration.

Reprinted with permission from Miekisch W, Schubert JK, Noeldge-Schomburg GFE. Diagnostic potential of breath analysis—focus on volatile organic compounds. Clinica Chimica Acta. 2004;347(1-2):25–39. Elsevier copyrights.

analysis, as they do not contain dead volume; thus the signal from biomarkers is expected to be two to three times higher compared to the mixed approach and have the lowest concentration of contaminants [4]. In this case, the exhaled CO_2 profile (Fig. 7.1) could help control the exact time for sampling, as it is person-to-person related. Regardless of the EBS procedure, researchers are also working on patient preparation before EBA, which includes various resting times and positions, exhalation of purified air before the EBS/EBA, wearing a nose clip, and/or holding their breath for a while. One popular technique is the wash-out breath maneuver (also known as lung wash-out). In this approach, subjects are asked to breathe synthetic air for a specific period before measurement (adjusted experimentally). Although this method seems to be the most effective for removing exogenous biomarkers from exhaled breath samples, the technical complications—such as being time-consuming—appear too high to implement routinely in clinical settings [4].

7.2 Heart failure

7.2.1 Definition and causes of heart failure

HF is not a distinct pathological condition but rather a clinical syndrome characterized by key symptoms such as breathlessness, swelling of the ankles, and fatigue. Physical signs, such as elevated jugular venous pressure, pulmonary crackles, or peripheral edema, may accompany these symptoms. The syndrome arises from structural and/or functional abnormalities of the heart, leading to increased intracardiac pressures and/or an insufficient cardiac output—either at rest or during physical exertion.

In most cases, the condition stems from left ventricular (LV) dysfunction, which can be systolic, diastolic, or a combination of both. Nevertheless, abnormalities affecting the heart valves, pericardium, and endocardium, as well as disorders of cardiac rhythm and conduction, may also cause or contribute to HF. HF may also result from right ventricular (RV) dysfunction, as RV mechanics and function become impaired under conditions of pressure or volume overload. Although the primary etiology of chronic RV failure is LV dysfunction-induced pulmonary hypertension, there are several other causes of RV dysfunction, including pulmonary hypertension of any other reason, arrhythmogenic RV cardiomyopathy (ARVC), or valve disease.

7.2.2 Types of heart failure

HF has traditionally been categorized into distinct phenotypes based on measurements of left ventricular ejection fraction (LVEF). This classification is based on early clinical trials, which demonstrated significantly improved outcomes in patients with LVEF ≤40% who received treatment.

Table 7.1 The New York Heart Association functional classification is based on the severity of symptoms and the level of physical activity.

Class I	No limitation of physical activity. Ordinary physical activity does not cause undue breathlessness, fatigue, or palpitations.
Class II	Slight limitation of physical activity. Comfortable at rest, but ordinary physical activity results in undue breathlessness, fatigue, or palpitations.
Class III	Marked limitation of physical activity. Comfortable at rest, but less than ordinary activity results in undue breathlessness, fatigue, or palpitations.
Class IV	Unable to carry on any physical activity without discomfort. Symptoms at rest can be present. If any physical activity is undertaken, discomfort is increased.

The current approach adopts the following classification of HF [5]:

- HF with reduced ejection fraction (HFrEF) is defined as LVEF ≤40%, reflecting a marked impairment in LV systolic function.
- HF with mildly reduced ejection fraction (HFmrEF) includes patients with an LVEF between 41% and 49%, indicating a modest decrease in systolic function.
- HF with preserved ejection fraction (HFpEF) refers to patients exhibiting symptoms and signs of HF, in combination with evidence of structural and/or functional cardiac abnormalities and/or elevated levels of natriuretic peptides, and an LVEF ≥50%.

This is just one way to classify HF. The nature of symptom onset can also categorize it as either chronic heart failure (CHF) or acute heart failure (AHF). CHF refers to patients whose symptoms are stable over time, whereas AHF involves a sudden or rapid onset. Decompensated HF describes a worsening of chronic symptoms, often requiring hospitalization or intravenous therapy. Another approach is to classify HF by symptom severity, typically using the NYHA functional classification (Table 7.1).

7.2.3 Epidemiology, etiology, and prognosis of heart failure

HF affects approximately 1%–2% of adults in Europe, with an overall incidence of about 3 per 1000 person-years and 5 per 1000 among adults. The true prevalence is likely underestimated due to the presence of undiagnosed cases. HF becomes more common with age, rising from ~1% in those under 55 to over 10% in people aged 70 or older. Approximately half of HF patients have HFrEF, while the remainder have HFpEF or HFmrEF. Data from outpatient registries, such as the European Society of Cardiology (ESC) Long-Term Registry, show 60% HFrEF, 24% HFmrEF, and 16% HFpEF. Slightly more than half of HF patients are women [6].

The etiology of HF varies according to geography. In Western-type and developed countries, coronary artery disease (CAD) and hypertension are predominant factors.

HF remains a significant cause of morbidity and mortality worldwide. According to the 2023 update of the ESC guidelines, mortality rates in HF vary based on the clinical presentation (HFrEF or HFpEF). Data from the ESC Heart Failure Long-Term Registry collected 2011–13 indicate that the 1-year mortality rate is approximately 23.6% for patients with acute HF and 6.4% for those with chronic HF (8.8% with HFrEF and 6.3% with HFpEF). When considering the combined endpoint of mortality or HF hospitalization within one year, rates are 36% for acute HF and 14.5% for chronic HF. More recent data on HF mortality—reflecting treatment practices primarily in line with the 2016 ESC guidelines—will be provided by the ESC Heart Failure III Registry, based on data collected between 2018 and 2020. Although these data precede the 2021 (updated in 2023) guidelines, the core recommendations have remained broadly consistent [6].

7.2.4 Diagnosis of heart failure

HF is defined by the presence of structural or functional cardiac abnormalities leading to impaired ventricular filling or ejection of blood. Consequently, cardiac imaging plays a pivotal role in its diagnosis. Echocardiography is considered the gold standard due to its widespread availability, noninvasive nature, and ability to comprehensively assess cardiac structure and function.

However, the clinical presentation of HF often includes nonspecific symptoms such as dyspnea, fatigue, and peripheral edema, which can be attributed to various other conditions. Given the high prevalence of these symptoms in the general population, the indiscriminate use of imaging modalities, such as echocardiography, for all suspected cases is neither practical nor cost-effective.

To address this, an integrated diagnostic approach is recommended. This involves initial clinical assessment, evaluation of risk factors, and the use of biomarkers to stratify patients before proceeding to imaging studies. Fig. 7.2 illustrates the recommended diagnostic algorithm for HF, emphasizing the stepwise evaluation process.

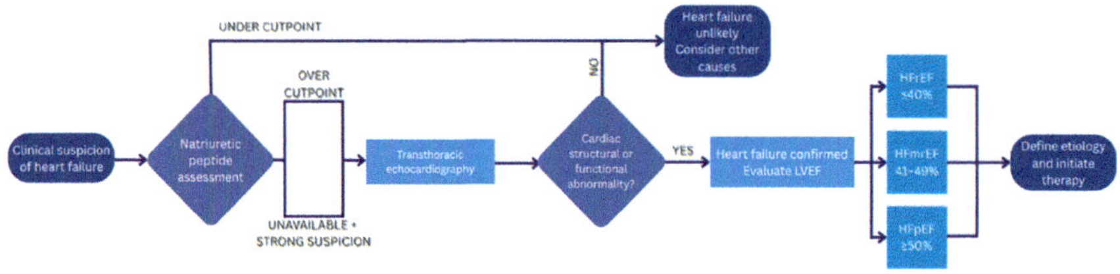

FIGURE 7.2

Diagnostic algorithm for heart failure.

Natriuretic peptides, specifically B-type natriuretic peptide (BNP) and N-terminal proBNP (NT-proBNP), play a crucial role in this diagnostic strategy. Elevated levels of these biomarkers reflect increased cardiac wall stress and are indicative of HF. Their role in the diagnosis and monitoring of HF is discussed in Chapter 5.4.3.

7.2.5 Heart failure treatment

The modern era of HF therapy truly began in the early 1990s, with the demonstration that angiotensin-converting enzyme inhibitors (ACEIs) could markedly improve survival. Over the ensuing decades, four classes of drugs have become foundational in the management of HFrEF: ACEIs/ARBs (or ARNIs), β-blockers, mineralocorticoid receptor antagonists (MRAs), and sodium-glucose cotransporter two inhibitors (SGLT2i). Sequential addition of these therapies, or replacement of ACEIs by ARNIs, has been shown to reduce mortality by approximately 15%–30% and HF hospitalizations by 20%–35%.

These data pertain specifically to HF with reduced ejection fraction. Historically, HFrEF has been the predominant focus of clinical trials—being long considered synonymous with "heart failure"—and untreated HFrEF carries the poorest prognosis; consequently, most evidence for disease-modifying therapies derives from this subgroup. The 2021 ESC heart failure guidelines, updated in 2023, recommend that all patients with HF should receive an SGLT2 inhibitor unless there is a clear contraindication. Other drug classes (ACEIs/ARBs/ARNIs, β-blockers, and MRAs) remain obligatory in HFrEF but are only recommended "should be considered" in HFmrEF.

In addition to pharmacological therapy, two device-based interventions—implantable cardioverter-defibrillator and cardiac resynchronization therapy (CRT)—have proven to improve outcomes in selected patients with HF. These device-based treatments complement guideline-directed medical therapy and have become key components in comprehensive HF management for patients who are appropriately selected.

7.2.6 Challenges in real-world management of heart failure

Despite significant advances in the treatment of HF, translating evidence from randomized clinical trials into routine clinical practice remains a considerable challenge. While trials have demonstrated substantial reductions in mortality and hospitalizations with guideline-recommended therapies, real-world data often show more modest improvements. For example, in the pre-ACEI era (1980s–early 1990s), 5-year survival for patients with newly diagnosed HF was only 50.4%, and 10-year survival was just 14.7% [7]. In comparison, a systematic review including 1.5 million ambulatory patients with CHF through 2018 reported pooled 5- and 10-year survival rates of 56.7% and 34.9%, respectively [8]. These figures demonstrate that, although outcomes have improved over time, real-world survival gains remain smaller than those expected based on clinical trial data.

Undoubtedly, a key driver of the gap between trial and clinical outcomes in HFrEF is that many patients do not receive all guideline-recommended therapies before the introduction of SGLT2i. Clinical trials and ESC guidelines recommended at least three foundational drug classes for HFrEF: a renin–angiotensin system inhibitor (ACEI, ARB, or ARNI), a β-blocker, and an MRA. In the CHECK-HF registry from the Netherlands (2013–16), which reflects the era when three drug classes were the standard of care, only 35.6% of HFrEF patients were treated with all three recommended therapies [9].

Unfortunately, real-world data on the use of all four drug classes (including SGLT2i) is still poorly available. Given the inertia in the adoption of even the first three therapies, it is reasonable to assume that comprehensive quadruple therapy remains even less common in everyday practice.

Norway is a bright spot in real-world HF management. Data from the Norwegian Heart Failure Registry show that, by 2023, 66% of HFrEF patients were receiving all four foundational drug classes (Fig. 7.3). This achievement reflects the success of Norway's nationwide network of specialized heart failure clinics but also underscores that such a model—supported by a wealthy, well-resourced health system—may be difficult to replicate elsewhere.

Looking ahead, noninvasive biomarker methods—such as analysis of VOCs in exhaled breath—offer great promise for improving HF care. These techniques could enable earlier detection of disease progression, real-time treatment monitoring, and

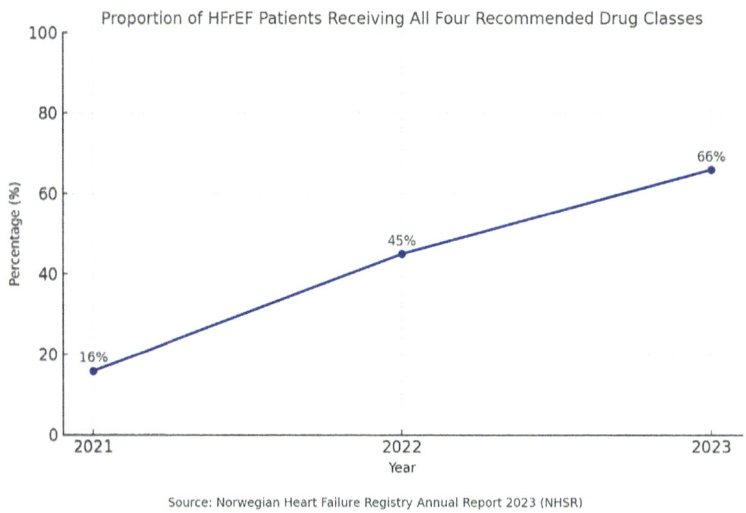

FIGURE 7.3

Use of Four Guideline-Directed Drug Classes among patients with HFrEF in the Norwegian Heart Failure Outpatient Clinic Network.

more personalized therapy adjustments. Wider adoption of breath-based biomarkers may help close the gap between trial results and everyday practice by guiding clinicians to optimize treatment more rapidly.

7.3 The common exhaled biomarkers in heart failure

7.3.1 Acetone

Acetone (2-propanone) is the primary ketone in exhaled human breath. It is generally detected in the ppb to ppm range. It is usually associated with diabetes studies [10], where the exhaled acetone level (EAL) is correlated to blood glucose levels and glycated hemoglobin (HbA1c); however, for cardiovascular disease diagnosis and monitoring, the EAL is also considered, and clinical trials are still ongoing, as elaborated in the following text.

7.3.2 Isoprene

Isoprene is one of the abundant endogenous VOCs in exhaled human breath. Apart from exhaled breath samples, isoprene has also been detected in blood [11], urine [12], and skin emanations [13]. Therefore in the last decades, exhaled isoprene levels (EIL) have drawn the attention of many researchers. An outstanding review on isoprene in human breath was delivered by Mochalski et al. [14] in 2023, including a historical overview of exhaled isoprene detection that began in 1960, when isoprene was first detected in breath [14]. The authors reviewed the recently published works in which EIL was the research subject. For example, King et al. [15] analyzed the EIL under nonsteady-state conditions, such as exercise, biking, breathing maneuvers, and position changes. The major conclusion from this work was that short rest periods after exercise may not be sufficient to achieve steady-state EIL, and thus considering exhaled isoprene as a biomarker is very challenging, as a wide concentration range needs to be established. Moreover, the unanswered question is where and how isoprene is produced in the body [14]. In 2001, Karl et al. [16] confirmed that EIL increases shortly after the start of exercise. However, the impact of sex, BMI, and metabolic condition has not been fully confirmed in the case of EIL [17]. Regarding cardiovascular diseases, McGrath et al. [18] analyzed breath samples collected from 12 patients with CHF and 31 healthy control subjects free from heart disorders. The CHF subjects were evaluated using the NYHA classification model, resulting in classifications of II and III. Breath was collected into a 60-L sample bag and further analyzed by the GC/MS technique (a Perkin-Elmer ATD400 coupled with an HP 5890 GC), including reference air samples. All samples were collected at 10:00 a.m. after subjects had been sitting at rest for 15 minutes. Isoprene was calculated as a concentration (nmol/L) and as a weight-adjusted production rate (pmol/min/kg). In turn, BIL in CHF subjects was significantly reduced ($P < 0.05$) compared to healthy subjects, 82(23) versus 168(20) pmol/min/kg [18]. Mendis et al. [19] analyzed the concentrations of pentane (exhaled pentane levels—EPL) and isoprene as potential biomarkers of acute myocardial infarction (AMI). The authors recruited 15 patients with stable angina and 15 healthy controls. Both groups were matched in terms of age, sex, smoking habits, hypertension, and cholesterol levels. The EPL and EIL were analyzed using the gas chromatography method. The results showed no significant difference in EPL between the AMI group and the control group, with values of 0.31 nmol/L and 0.36 nmol/L, respectively. On the other hand, EIL was higher ($p < 0.01$) in the AMI group, at 11.4 nmol/L and 7.1 nmol/L. However, there was no difference in EIL or EPL between the control group and the group with stable angina. Interestingly, Neupabe et al. [20] observed that EIL rises during hypoglycemic in T1DM subjects. The observation was made on EIL samples collected from eight female nonsmoking subjects with type 1 diabetes mellitus (T1DM). The average duration of diabetes was 23±7 years (no statin was used). The analysis was conducted using a single-blinded, computer code-randomized crossover design. For breath collection, subjects held their breath for 3 seconds, partially exhaled, and breathed into a 1.1-L breath bag. A soft-ionization mass spectrometry (V&F AIRSENSE Compact Ion Molecule Reaction Mass Spectrometer) was used as a reference device. The authors concluded that it is unclear how hypoglycemic could increase isoprene. Despite being one of the most common VOCs in human breath, the source of endogenous isoprene remains undetermined [20].

7.3.3 Pentane

EPL has been considered a reliable biomarker of lipid peroxidation and oxidative stress. The linkage between EPLs was a subject of research conducted by Sobotka et al. [21]. The authors recruited 13 patients with congestive HF and the same number of age-matched subjects to serve as a healthy control group. Neither the study participants nor the researchers took ACEIs at the time of the study. Since EILs are unaffected by food intake, participants were not required to fast before exhaled breath collection. The breath samples consisted of 50 mL of end-expiratory air and were analyzed using the GC (GC 3600, Varian Instruments) technique. At the time of the study, CHF patients were classified according to the NYHA classification, with associated classifications of II, III, and IV; however, the number of subjects in each specific group changed over the course of the study. The analysis showed that patients with congestive HF excreted significantly higher ($p < 0.005$) EPL than the control subjects, that is, 5.7 ± 2.1 nmol/L and 3.6 ± 1.2 nmol/L, respectively. Additionally, EPL was correlated with the daily dose of captopril, a sulfhydryl-containing drug. Patients treated with captopril also had significantly higher ($p < 0.05$) excretion of EPL than control patients, that is, 4.7 ± 1.3 nmol/L. The dose of captopril therapy was considerably lower before ($p < 0.01$) and after ($p < 0.02$) nonsulfhydryl inhibitor therapy [21]. Tranchito et al. [22] analyzed EPL from a cohort of 50 patients undergoing cardiopulmonary exercise testing (CPET). The samples were measured by SIFT-MS, where 21 VOCs, including pentane, were correlated with minute ventilation/carbon dioxide production (VE/VCO_2). The results showed that patients with a VE/VCO_2 above the median (<35.5) had significantly higher initial resting pentane levels (15.95 ppb vs 11.68 ppb, $P = 0.049$). The baseline and post-exercise pentane were similar between those with higher versus lower peak VO_2 [22]. In 55 patients admitted with acute decompensated heart failure (ADHF), the authors measured exhaled breath acetone and pentane levels upon admission and after 48 hours of diuresis. In a separate cohort of 51 cardiac patients undergoing CPET, the authors measured exhaled breath acetone and pentane levels before and at peak exercise. In the ADHF cohort, admission acetone levels correlated with lower LVEF($r = -0.297$, $p = 0.035$). Greater weight loss with diuretic therapy was associated with a greater reduction in both acetone levels ($r = -0.398$, $p = 0.003$) and pentane levels ($r = -0.309$, $p = 0.021$). In patients with above-median weight loss (≥4.5 kg), a significantly greater percentage reduction was observed in acetone (59% reduction vs 7% increase, $p < 0.001$) and pentane (23% reduction vs 2% increase, $p = 0.008$). In the CPET cohort, admission acetone and pentane levels correlated with higher VE/VCO_2 ($r = 0.39$, $p = 0.005$) and ($r = 0.035$, $p = 0.014$). However, there were no significant correlations between baseline or peak exercise acetone and pentane levels and peak VO_2. In longitudinal follow-up with a median duration of 33 months, patients with elevated exhaled acetone and pentane levels experienced higher composite adverse events of death, ventricular assist device implantation, or orthotopic heart transplantation in patients admitted with ADHF [22].

7.3.4 Nitric oxide

Nitric oxide (NO) is produced mainly by endothelial cells lining blood vessels, and reduced NO production is a hallmark of endothelial dysfunction associated with cardiovascular disease, including HF. Indeed, Adachi et al. [23] found reduced exhaled NO concentration (measured by chemiluminescence method) in HF versus healthy subjects both at rest (4.0 ± 2.2 ppb vs 10.5 ± 6.2 ppb, respectively; $p < 0.05$) and at peak exercise (75.3 ± 43.4 nL/min vs 512.9 ± 253.6 nL/min, respectively; $p < 0.01$)—however, Hare et al. [24] measured the mean mixed expired NO content of a vital-capacity breath using chemiluminescence in patients with compensated HF ($n = 30$), decompensated HF ($n = 7$), and in normal control subjects ($n = 90$) with corresponding pulmonary artery pressures. Exhaled nitrogen oxide (eNO) was higher in patients with HF than in control subjects (9.9 ± 1.1 ppb vs. 6.2 ± 0.4 ppb, $p = 0.002$) and was inversely correlated with PAP. Patients with decompensated HF exhibited even higher levels of eNO (20.4 ± 6.2 ppb) and PAP, but there was a loss of the inverse relationship between these two variables. During therapy (7.3 ± 6 days) with sodium nitroprusside and diuresis, hemodynamics improved, eNO concentrations fell (11.2 ± 1.2 ppb), and the relationship between eNO and PAP was restored. Thus the association between HF and eNO appears to be complex, and further studies are needed to clarify its nature.

7.3.5 Others

Carbon monoxide (CO) is a gas produced in the body as the byproduct of heme metabolism. It was reported to be both cytoprotective, as a signaling mediator, and detrimental, as an agent promoting oxidative stress and endothelial dysfunction. As CO equilibrates quickly across the alveolar-capillary membrane, exhaled CO (eCO) reflects endogenous CO production. Recently, Framingham Heart Study investigators have reported that eCO is directly associated with CV risk factors [25] and subclinical CV disease [26]. Recently, Tun et al. [27] showed that over a median follow-up of 18 years, 309 participants (of 3521 participants) developed HF. After multivariable adjustment, higher eCO was associated with higher risk of HF (hazards ratio per SD increment: 1.39; 95% CI, 1.19–1.62 [P <0.001]) and with higher risk of HFrEF (N = 144 events; hazard ratio per SD increment in eCO: 1.43; 95% CI, 1.15–1.77 [P = 0.001]). Thus, while data suggest that higher eCO is associated with increased CV risk and incident HF, no data are available on eCO in patients with established HF, nor on its response to HF exacerbation or therapy.

7.4 Noninvasive diagnostic methods in heart failure

Noninvasive methods play a pivotal role in the diagnosis and monitoring of HF, allowing for the assessment of cardiac structure, function, and functional reserve without the risks associated with invasive procedures.

7.4.1 Imaging methods

Echocardiography remains the cornerstone and most widely available imaging modality in patients with HF. It provides accurate measurement of LVEF, chamber volumes, regional wall motion abnormalities, and diastolic dysfunction using Doppler indices. Echocardiography also allows for the assessment of proper ventricular function, pulmonary artery systolic pressure, and coexisting valvular disease. Significantly, it contributes to identifying the underlying etiology of HF in a given patient, such as ischemic versus nonischemic cardiomyopathy or valvular pathology. Due to its availability, safety, and repeatability, echocardiography is considered the first-line diagnostic and follow-up tool in HF.

Nevertheless, this technique has several limitations. Image quality can be significantly impaired in patients with advanced chronic obstructive pulmonary disease, obesity, or chest wall abnormalities, resulting in suboptimal acoustic windows. Its accuracy is also dependent on operator expertise, with variability between observers. Moreover, echocardiography provides only indirect estimates of filling pressures and pulmonary artery pressures, which may be less reliable compared with invasive measurements. In patients with complex congenital or structural heart disease, echocardiography alone may not be sufficient, and complementary imaging modalities such as cardiac magnetic resonance (CMR) are often required.

CMR is considered the gold standard for the quantitative assessment of chamber volumes, myocardial mass, and ejection fraction. It is especially valuable in patients with suboptimal echocardiographic windows. With its high spatial resolution and advanced tissue characterization techniques, CMR allows for differentiation between ischemic and nonischemic cardiomyopathies, assessment of fibrosis and infiltration, and monitoring of treatment response. It is also instrumental in detecting infiltrative and storage diseases, such as amyloidosis or Anderson–Fabry disease, which are often challenging to diagnose with other imaging modalities.

Beyond diagnostic purposes, CMR can support the selection of patients for CRT. Scar imaging with late gadolinium enhancement helps identify the extent and location of myocardial fibrosis, which is associated with a response to CRT. Patients with extensive scarring, particularly in the posterolateral LV wall, may have less benefit, whereas the presence of viable myocardium in target regions predicts a more favorable outcome. Moreover, CMR can provide detailed information on LV desynchrony, although this application is not yet standardized in clinical practice. Thus, while echocardiography remains the primary imaging modality for CRT eligibility, CMR may refine patient selection and improve the prediction of therapeutic response [28].

Despite these strengths, CMR has several significant limitations. It is contraindicated in patients with certain metallic implants, and patients with non-MRI-conditional devices are generally excluded from undergoing CMR. Arrhythmias, particularly atrial fibrillation or frequent ectopy, can reduce temporal resolution and impair the accuracy of functional measurements. Claustrophobia and the need for prolonged breath-holding during image acquisition may limit feasibility in some patients. Furthermore, gadolinium-based contrast agents, often required for tissue characterization, are contraindicated in advanced renal dysfunction due to the risk of nephrogenic systemic fibrosis. Finally, CMR is less widely available than echocardiography, requires longer examination times, and is associated with higher costs, which may restrict its use in routine practice.

Computed tomography (CT), although less commonly used in the routine work-up of HF, can provide valuable complementary information. It is particularly constructive in patients with inconclusive echocardiographic findings, in the evaluation of complex valvular morphology or prosthetic valves, and in assessing extracardiac structures that may contribute to HF symptoms, such as aortic or pericardial disease.

Nuclear imaging techniques, including single-photon emission computed tomography (SPECT) and positron emission tomography (PET), play an essential role in the functional evaluation of HF. They provide insights into myocardial perfusion and viability, enabling the identification of hibernating myocardium that may benefit from revascularization. PET tracers also allow the assessment of inflammatory and infiltrative cardiomyopathies, including sarcoidosis and amyloidosis. In recent years, targeted nuclear imaging using radiolabeled compounds—-such as bone-seeking tracers for transthyretin amyloidosis or antibody-based scintigraphy—has expanded diagnostic possibilities, enabling earlier and more specific detection of infiltrative heart diseases.

7.4.2 Functional tests

Complementary to imaging are functional tests, which provide crucial information on exercise tolerance, ventilatory efficiency, and global cardiopulmonary performance. Although standard exercise testing on a treadmill or cycle ergometer has limited diagnostic value in HF, it remains helpful in assessing overall functional capacity, blood pressure response, and chronotropic competence and for unmasking exercise-induced arrhythmias.

CPET represents the gold standard for functional and prognostic evaluation in HF. By integrating gas exchange analysis with exercise parameters, CPET yields objective markers of disease severity. Among them, peak VO_2 is the most established index of aerobic capacity and a cornerstone criterion in heart transplantation listing (<14 mL/kg/min, or <12 mL/kg/min in patients receiving β-blockers). The VE/VCO_2 slope serves as an additional, powerful prognostic indicator, reflecting ventilatory efficiency and correlating with adverse outcomes. Together, these parameters allow accurate risk stratification and provide insight into treatment response over time.

In daily clinical practice, the 6-minute walk test (6MWT) remains a simple, safe, and widely applied alternative. Although less precise than CPET, the 6MWT captures functional status, symptom burden, and longitudinal changes in exercise tolerance. Distances below 300 m are generally associated with a poor prognosis, and serial testing is particularly valuable for monitoring therapy effectiveness or disease progression in advanced HF.

With technological advances, noninvasive diagnostic tools are enabling increasingly precise evaluations of cardiac morphology, function, prognosis, and therapeutic efficacy. Their safety and repeatability make them a cornerstone of modern clinical practice in HF management. Nevertheless, a clear need remains for the development of novel diagnostic approaches that are simpler, more widely accessible, and less dependent on costly equipment, such as CMR, or on highly specialized expertise, as required for echocardiography. This is particularly important in light of the growing number of patients with HF, a trend expected to accelerate with the aging of the global population. In this context, methods based on innovative biomarker analysis—for instance, the characterization of exhaled breath—hold considerable promise as affordable and scalable tools for early detection and longitudinal monitoring of HF.

7.4.3 Natriuretic peptides

Natriuretic peptides—B-type natriuretic peptide (BNP) and its inactive cleavage fragment, N-terminal proBNP (NT-proBNP), are hormones secreted primarily by ventricular cardiomyocytes in response to increased wall stress (volume and pressure overload). Their primary physiological actions include promoting natriuresis, diuresis, vasodilation, and inhibition of the renin–angiotensin–aldosterone system (RAAS) and the sympathetic nervous system. Compared to BNP, NT-proBNP has a longer half-life and is more stable in plasma, making it particularly useful for routine clinical testing. Elevated plasma concentrations of BNP and NT-proBNP are observed in a wide range of clinical conditions, most notably HF, acute coronary syndromes, tachyarrhythmias, valvular heart disease, pulmonary hypertension, and chronic kidney disease. A comprehensive overview of conditions associated with elevated natriuretic peptide levels is provided in Table 7.2.

In clinical practice, BNP and NT-proBNP are the most established circulating biomarkers in cardiology. They serve as essential diagnostic tools: elevated levels

Table 7.2 Causes of elevated concentrations of natriuretic peptide levels (from 2016 ESC guidelines for HF).

Cardiac	– Heart failure – Acute coronary syndromes – Pulmonary embolism – Myocarditis – Left ventricular hypertrophy – Hypertrophic or restrictive cardiomyopathy – Valvular heart disease – Congenital heart disease – Atrial and ventricular tachyarrhythmias – Heart contusion – Cardioversion, CD shock – Surgical procedures involving the heart – Pulmonary hypertension
Noncardiac	– Advanced age – Ischemic stroke – Subarchnoid hemorrhage – Renal dysfunction – Liver dysfunction – Paraneoplastic syndrome – Chronic obstructive pulmonary disease – Severe infections (including pneumonia and sepsis) – Severe burns – Anemia – Severe metabolic and hormone abnormalities

strongly support the diagnosis of HF, particularly in patients presenting with dyspnea of unclear origin. However, increased concentrations are not specific to HF and may also be seen in other conditions such as renal dysfunction, pulmonary hypertension, or acute coronary syndromes. Conversely, low BNP or NT-proBNP levels have a high negative predictive value, making the diagnosis of HF very unlikely. For CHF, the ESC 2021 guidelines recommend cutoff values of BNP <35 pg/mL and NT-proBNP <125 pg/mL. These thresholds are characterized by high sensitivity (≈85%–95%), meaning that almost all patients with HF have values above these levels, while specificity is moderate (≈40%–60%), reflecting that elevated levels may also occur in non-HF conditions. The NICE guidelines in the United Kingdom use a higher threshold of NT-proBNP <400 pg/mL, with similarly high sensitivity (≈90%–97%) but relatively low specificity (≈40%–50%).

They are instrumental in patients presenting with acute dyspnea, where low natriuretic peptide levels can reliably rule out acute HF. For BNP, values below 100 pg/mL have a sensitivity of about 90% and specificity of 76%, making acute HF very unlikely. Conversely, levels above 400 pg/mL reach a specificity of about 91%, supporting the diagnosis (rule-in). For NT-proBNP, a threshold of <300 pg/mL yields a sensitivity of 99% and specificity of 68%, effectively excluding acute HF. Rule-in, on the other hand, relies on age-specific cutoffs: ≥450 pg/mL for patients under 50 years, ≥900 pg/mL for those aged 50–75 years, and ≥1800 pg/mL for patients older than 75 years. At these thresholds, diagnostic performance reaches approximately 90% sensitivity and 86% specificity [29,30], enabling clinicians to differentiate acute HF from other causes of dyspnea.

Although natriuretic peptide levels, particularly NT-proBNP, rise with worsening HF, their correlation with the NYHA functional class is only modest. This reflects the subjective nature of the NYHA classification, which relies on patient-reported symptoms and clinician judgment, and may be confounded by comorbidities, age, and overall physical activity level. In contrast, NT-proBNP provides an objective measure of cardiac wall stress and has consistently demonstrated substantial prognostic value, independent of NYHA class, across multiple studies [30,31]. In this context, the use of breath biomarkers appears promising, as compounds such as acetone have been shown to correlate not only with NT-proBNP levels but also with symptom burden and prognosis.

7.5 Current challenges in exhaled breath analysis, including registered clinical trials in cardiovascular diseases

The current challenges in EBA are similar to those of other medical methods, namely the validation in clinical trials that the developed EBA analysis, focused on specific exhaled biomarkers, can be practically applied in daily clinical practice, not only under laboratory conditions.

Recently, Biagini et al. [32] reported the results obtained during the AEOLUS pilot study, in which EBA was combined with CPET and echocardiographic examination for monitoring HF cases. The study was conducted to analyze the VOC profiles between the two groups. In the study, 10 patients with a prior clinical diagnosis of HF with reduced LVEF and 15 patients with CVD risk factors such as hypertension, type 2 diabetes mellitus, or chronic ischemic heart disease and no history of HF were enrolled. The inclusion criteria were based on the NT-proBNP level, that is, below the <125 ng/L threshold value and LVEF above 50%. No loop diuretic therapy was also allowed. The list of exclusion criteria parameters consists acute myocardial ischemia, more than moderate left-sided valvular disease at rest or during exercise, atrial fibrillation, arrhythmias (including not well controlled by antiarrhythmic drugs), or implantable defibrillator/pacemaker, inability to complete the exercise, lower extremity artery disease, or inadequate acoustic windows. All subjects underwent a comprehensive clinical assessment, including blood tests, a 12-lead ECG, spirometry verification (subjects with moderate or greater airflow obstruction were excluded), and no ultrasound evaluation. For NTME-GC-MS analysis, 50 mL of mixed samples were collected at a rate of 5 mL/min from enrolled patients at rest and during the CPET-ESE protocol [32] using a sampler device. In parallel, 50 mL of ambient air was also collected at the same speed to analyze the background contamination. After the analysis, the NTDs were rinsed with methanol to remove any compounds that may have been adsorbed during the sampling process. A 7890 GC (Agilent Technologies, USA) and a 7010 triple quadrupole GC/MS (Agilent Technologies, USA) detection system were used for VOC analysis. The specific parameters of the devices and condition parameters for NTDs are given in the reference paper. SIFT-MS measurements were conducted using a Voice200 Ultra SIFT-MS instrument (Syft Technologies, New Zealand), where breath-by-breath acetone and isoprene were detected. The data analysis used the T-test and the Mann-Whitney test to evaluate potential differences between the control and HF groups. Detailed information about the data analysis is given in the reference; however, the obtained results suggest that during rest, pentane (2043 vs 948 ppbv) and acetone (417 vs 184 ppbv) were significantly different for HF and control, with a p-value < 0.01, while during the CPET test, only acetone (393 vs 171 ppbv) reached $p = 0.01$, and isoprene (68 vs 115 ppbv) was $p = 0.04$. Other compounds, such as dimethyl sulfide, carbon disulfide, toluene, hexanone, heptanone, nonanone, butanone, butanedione, and dimethyl disulfide, have not shown significant differences in either test. Additionally, the authors used a capnogram to calculate the number of end-tidal points, frequency, average, and standard deviation of the acetone and isoprene levels collected using the SIFT-MS method. The increased value of acetone for HF patients can be explained by the fact that the recruited patients were characterized by a reduced LVEF associated with fluid accumulation in the intravascular compartment and the interstitial space. Therefore it can be assumed that HF patients exhibit altered myocardial and systemic metabolism, resulting in increased blood ketone body levels. The correlation between exhaled acetone (analyzed by the NTME-GC-MS analysis) and plasma NT-proBNP levels showed a good correlation ($r = 0.646$, $p < 0.001$) in both groups, that is, HF and control subjects. Interestingly, exhaled acetone analysis is also considered for blood glucose monitoring [10]; however, the EAL differ between diabetic and nondiabetic subjects. In this study, all subjects fasted overnight, which is a factor that should be considered. On the other hand, the EPL did not correlate with plasma NT-proBNP, suggesting a poor relationship with the HF severity. Although the increased EPL has been observed in HF patients due to enhanced reactive oxygen species production, more clinical data are required to confirm the hypothesis that EPL can be used as an EAL for screening purposes. It is essential to emphasize that online methods, which enable real-time measurements and provide results quickly, are preferable for screening tests compared to offline methods that require time, consumables, and a well-equipped laboratory. Therefore the SIFT-MS method seems to have a higher potential for implementation in daily clinical practice. Nevertheless, the AEOLUS study paves the way for further analysis on a higher cohort of HF patients to validate the positive correlation between EAL and NT-proBNP levels in HF patients and the high CVD-risk control group.

Gouzi et al. [33] attempted to address this issue by systematically reviewing databases (1991–2020) in which EAL analysis was used for HF diagnosis. The authors identified 971 studies involving 1842 subjects, comprising 833 healthy controls and 1009 with HF. The mean EAL was 1.89 times higher in HF patients versus HC (782 [531–1032] vs 413 [347–478] ppbv; $p < 0.001$). Interestingly, a vital heterogeneity of breath acetone concentrations between studies in healthy controls ($p < 0.01$) was observed, although it was not possible to provide a standard range of EAL [33]. The authors observed that fasting acetone concentrations were higher than those on an uncontrolled diet (703 (581–825) ppbv vs 387 (320–454) ppbv; $p < 0.001$). From an analytical perspective, online measurement systems (IMR-MS, SIFT-MS, and PTR-MS) yielded higher values than offline systems (GC and GC-MS), specifically 576 (479–623) ppbv vs 382 (313–452) ppb; $p < 0.001$. One important observation was also made: alveolar samples and end-tidal samples did not differ significantly from total breath sampling concentrations (377 (306–448) ppbv vs. 488 (249–727) ppbv; $p = 0.38$). Another observation that should be highlighted is that EAL was higher when measurements were obtained in AHF patients than in studies including stable HF. In addition, between 2020 and 2025, new papers have been published investigating EAL for HF diagnosis, although their impact has been limited due to the small number of subjects enrolled. Also, the number of subjects involved in the studies and the poor quality of the control subjects' selection are significant limitations of the analyzed studies.

However, it is essential to understand that variations such as fasting/no-fasting stage, sampling type, and analytical method (offline vs online) significantly impact the reported exhaled breath levels. Therefore, the metaanalysis should be conducted by comparing the same methodology on a sufficient cohort of subjects with proper heterogeneity, as it is undertaken for drug clinical trials. In 2009, Schwarz et al. [34] conducted studies to determine the variations of EAL with age, gender, and body mass index (BMI). Mixed expiratory exhaled breath was sampled using Tedlar bags and analyzed using the PTR-MS method. Two hundred forty-three adult volunteers, who were not diabetic, were recruited. The EAL was in the 177–2441 ppb range. The authors did not see a statistically significant shift between the distributions of EAL in the breath of males and females in the fasted group and in the group with dietary control. Aging was associated with a slight increase in acetone in the fasted females; in males, the increase was not statistically significant. No correlation was found between the acetone levels and BMI in adults [34].

The major challenge so far is to carry out a multi-center study, and choosing a single analytical method (e.g., gas chromatography (GC) coupled with mass spectrometry (MS) as GC-MS, ion-molecule reaction mass spectrometry (IMR-MS), selected ion flow-tube mass spectrometry (SIFT-MS), proton transfer time of flight-mass spectrometry (PTR-TOF-MS), electronic noses (e-nose), and portable exhaled breath analyzers (PEBA)) for such tests. Physiological factors such as age, sex, weight, food consumption, and habits (including active and past tobacco use) must also be considered. It is clear that each method, as mentioned above, has its pros and cons, and the researchers preferentially use the technique that they know best and have access to. These limitations reduce the practical application of EAL. In our opinion, a worldwide consensus must be reached to overcome these limitations and initiate a comprehensive research program that enables answering the question: Is it possible to detect and monitor the progress of HF using EBA, with special emphasis on the VOCs present in exhaled breath? Another issue is related to the value of the cutoff EAL, since Marcondes-Braga et al. [35] suggest that the EAL cutoff value could be 496 ppbv (sensitivity 83%, specificity 100%, accuracy 86% measured with samples collected from 89 and 20 for HF and HC subjects, respectively). In contrast, by changing the detection method, the same group suggests that the EAL cutoff value could be > 513 ppbv [35].

Interestingly, exhaled breath acetone concentration measurements have also been investigated for diabetes management and diagnosis [36,37], raising an obvious question about the potential use of EAL in these two diseases. So far, this issue remains unresolved; however, Marcondes-Braga et al. [37] reported that EAL was not influenced by the presence of diabetes in the GENIUS-HF study conducted by the authors. In detail, between August 2012 and December 2014, 695 patients were enrolled in the GENIUS-HF study. Most patients were characterized by hypertensive or ischemic etiology, NYHA class I/II with a mean LVEF of 32% (± 8.6%), and median BNP levels of 149 (54–355) pg/mL. Twenty percent of patients were taking oral hypoglycaemic drugs, but none of them were taking SGLT2i. The authors observed that EAL levels reflected clinical changes of HF in patients with chronic stable disease, that is, a significant correlation between EAL and right HF symptoms (such as dyspnea, orthopnoea, paroxysmal nocturnal dyspnea, jugular venous distension, pulmonary rales, peripheral oedema, third heart sound, hepatojugular reflux, capillary filling time higher than 3 seconds, ascites, and hepatomegaly) where EAL increases as the number of HF symptoms/signs increases [37]. The median exhaled breath acetone concentration was 0.6

(0.3–1.2) μg/L. Furthermore, EAL values > 1.2 μg/L indicated a significantly worse prognosis (log-rank test, P <0.001). They can be used as an independent predictor of cardiac ($P = 0.011$) and overall ($P = 0.010$) mortality in the cohort enrolled in the GENIUS-HF study in Brazil.

7.6 Conclusions

EBA, in general, and especially for cardiovascular diseases, requires accurate and validated methods, devices, and protocols. Regardless of the number of compounds in exhaled breath that will be analyzed to determine whether a patient has a particular disorder, whether treatment is needed, and, if so, whether treatment should be adjusted, the noninvasive character of EBA analysis paves the way for its global use. However, it can also be done only after global research is conducted in many centers with a large number of patients enrolled. As is becoming increasingly appreciated, many CVDs have subtypes, and thus the variability of exhaled breath biomarkers may also be observed. Typically, a new diagnostic method is verified against a reference diagnosis to calculate clinical sensitivity. In the case of HF, the gold standard is based on the NT-proBNP blood test. For a perfect diagnostic biomarker, a test should result in 100% sensitivity and 100% specificity. However, no biomarker test has been developed with such results, and EBA also seems to suffer from this limitation. EBA could be considered a method for measuring disease progression, including the worsening of treatment. However, a specific set of biomarkers has been established first, including those mentioned above and the novel ones that can appear during the cohort studies.

Furthermore, effective and clear communication is crucial for the efficient translation of EBA into the approved medical method, with special emphasis on interdisciplinary cooperation throughout all phases, including discovery, validation, hypothesis formulation, clinical trials, and reporting. Transparency of research progress and scientific evidence blaze a trail for regulatory agencies (e.g., FDA) to approve the proposed method.

References

[1] Breath biopsy—world-leading breath analysis (n.d.). https://www.owlstonemedical.com/science-technology/breath-biopsy/ (accessed September 30, 2025).

[2] Morrow DA, De Lemos JA. Benchmarks for the assessment of novel cardiovascular biomarkers. Circulation 2007;115:949–52. Available from https://doi.org/10.1161/CIRCULATIONAHA.106.683110.

[3] Herbig J, Beauchamp J. Towards standardization in the analysis of breath gas volatiles. Journal of Breath Research 2014;8:037101. Available from https://doi.org/10.1088/1752-7155/8/3/037101.

[4] Miekisch W, Schubert JK, Noeldge-Schomburg GFE. Diagnostic potential of breath analysis—focus on volatile organic compounds. Clinica Chimica Acta 2004;347:25–39. Available from https://doi.org/10.1016/J.CCCN.2004.04.023.

[5] McDonagh TA, Metra M, Adamo M, Gardner RS, Baumbach A, Böhm M, et al. 2023 focused update of the 2021 ESC guidelines for the diagnosis and treatment of acute and chronic heart failure: developed by the task force for the diagnosis and treatment of acute and chronic heart failure of the European Society of Cardiology (ESC) with the special contribution of the Heart Failure Association (HFA) of the ESC. European Heart Journal 2023;44:3627–39. Available from https://doi.org/10.1093/EURHEARTJ/EHAD195.

[6] Crespo-Leiro MG, Anker SD, Maggioni AP, Coats AJ, Filippatos G, Ruschitzka F, et al. European Society of Cardiology Heart Failure Long-Term Registry (ESC-HF-LT): 1-year follow-up outcomes and differences across regions. European Journal of Heart Failure: Journal of the Working Group on Heart Failure of the European Society of Cardiology 2016;18:613–25. Available from https://doi.org/10.1002/EJHF.566.

[7] Vaughan DE, Lamas GA, Pfeffer MA. Role of left ventricular dysfunction in neurohumoral activation in the recovery phase of anterior wall acute myocardial infarction. The American Journal of Cardiology 1990;66:529–32. Available from https://doi.org/10.1016/0002-9149(90)90476-H.

[8] Jones NR, Roalfe AK, Adoki I, Hobbs FDR, Taylor CJ. Survival of patients with chronic heart failure in the community: a systematic review and meta-analysis. European Journal of Heart Failure: Journal of the Working Group on Heart Failure of the European Society of Cardiology 2019;21:1306–25. Available from https://doi.org/10.1002/EJHF.1594.

[9] Linssen GCM, Veenis JF, Brunner-La Rocca HP, van Pol PEJ, Engelen DJM, van Tooren RM, et al. Differences in guideline-recommended heart failure medication between Dutch heart failure clinics: an analysis of the CHECK-HF registry. Netherlands Heart Journal 2020;28:334–44. Available from https://doi.org/10.1007/S12471-020-01421-1/TABLES/3.

[10] Rydosz A. Exhaled breath analysis : current status, challenges and future perspectives. 2025. p. 200.

[11] Cailleux A, Cogny M, Allain P. Blood isoprene concentrations in humans and in some animal species. Biochemical Medicine and Metabolic Biology 1992;47:157–60. Available from https://doi.org/10.1016/0885-4505(92)90019-U.

[12] Buckley LA, Coleman DP, Burgess JP, Thomas BF, Burka LT, Jeffcoat AR. Identification of urinary metabolites of isoprene in rats and comparison with mouse urinary metabolites. Drug Metabolism and Disposition 1999;27:846–52. Available from https://doi.org/10.1016/S0090-9556(24)11819-3.

[13] Mochalski P, King J, Unterkofler K, Hinterhuber H, Amann A. Emission rates of selected volatile organic compounds from skin of healthy volunteers. Journal of Chromatography. B, Analytical Technologies in the Biomedical and Life Sciences 2014;959:62. Available from https://doi.org/10.1016/J.JCHROMB.2014.04.006.

[14] Mochalski P, King J, Mayhew CA, Unterkofler K. A review on isoprene in human breath. Journal of Breath Research 2023;17:037101. Available from https://doi.org/10.1088/1752-7163/ACC964.

[15] King J, Koc H, Unterkofler K, Mochalski P, Kupferthaler A, Teschl G, et al. Physiological modeling of isoprene dynamics in exhaled breath. Journal of

Theoretical Biology 2010;267:626–37. Available from https://doi.org/10.1016/J.JTBI.2010.09.028.

[16] Karl T, Prazeller P, Mayr D, Jordan A, Rieder J, Fall R, et al. Human breath isoprene and its relation to blood cholesterol levels: new measurements and modeling. Journal of Applied Physiology 2001;91:762–70. Available from https://doi.org/10.1152/JAPPL.2001.91.2.762.

[17] Sukul P, Richter A, Junghanss C, Schubert JK, Miekisch W. Origin of breath isoprene in humans is revealed via multi-omic investigations. Communications Biology 2023;6:1–12. Available from https://doi.org/10.1038/S42003-023-05384-Y;TECHMETA.

[18] McGrath LT, Patrick P, Silke B. Breath isoprene in patients with heart failure. European Journal of Heart Failure: Journal of the Working Group on Heart Failure of the European Society of Cardiology 2001;3:423–7. Available from https://doi.org/10.1016/S1388-9842(01)00128-3.

[19] Mendis S, Sobotka PA, Euler DE. Expired hydrocarbons in patients with acute myocardial infarction. Free Radical Research 1995;23:117–22. Available from https://doi.org/10.3109/10715769509064026.

[20] Neupane S, Peverall R, Richmond G, Blaikie TPJ, Taylor D, Hancock G, et al. Exhaled breath isoprene rises during hypoglycemia in type 1 diabetes. Diabetes Care 2016;39:e97–8. Available from https://doi.org/10.2337/DC16-0461.

[21] Sobotka PA, Brottman MD, Weitz Z, Birnbaum AJ, Skosey JL, Zarling EJ. Elevated breath pentane in heart failure reduced by free radical scavenger. Free Radical Biology and Medicine 1993;14:643–7. Available from https://doi.org/10.1016/0891-5849(93)90145-K.

[22] Tranchito L, Albert CL, Gul Z, Cikach F, Grove D, Dweik R, et al. Exhaled acetone and pentane in patients undergoing cardiopulmonary exercise testing. Journal of Cardiac Failure 2017;23:S31. Available from https://doi.org/10.1016/j.cardfail.2017.07.080.

[23] Adachi H, Nguyen PH, Belardinelli R, Hunter D, Sung T, Wasserman K. Nitric oxide production during exercise in chronic heart failure. American Heart Journal 1997;134:196–202. Available from https://doi.org/10.1016/S0002-8703(97)70124-8.

[24] Hare JM, Nguyen GC, Massaro AF, Drazen JM, Stevenson LW, Colucci WS, et al. Exhaled nitric oxide: a marker of pulmonary hemodynamics in heart failure. Journal of the American College of Cardiology 2002;40:1114–9. Available from https://doi.org/10.1016/S0735-1097(02)02117-4.

[25] Cheng S, Lyass A, Massaro JM, O'Connor GT, Keaney JF, Vasan RS. Exhaled carbon monoxide and risk of metabolic syndrome and cardiovascular disease in the community. Circulation 2010;122:1470–7. Available from https://doi.org/10.1161/CIRCULATIONAHA.110.941013.

[26] Cheng S, Enserro D, Xanthakis V, Sullivan LM, Murabito JM, Benjamin EJ, et al. Association of exhaled carbon monoxide with subclinical cardiovascular disease and their conjoint impact on the incidence of cardiovascular outcomes. European Heart Journal 2014;35:2980. Available from https://doi.org/10.1093/EURHEARTJ/EHU052.

[27] Tun B, Ehrbar R, Short M, Cheng S, Vasan RS, Xanthakis V. Association of exhaled carbon monoxide with ideal cardiovascular health, circulating biomarkers, and incidence of heart failure in the framingham offspring study. Journal of the American Heart Association: Cardiovascular and Cerebrovascular Disease 2020;9, e016762. https://doi.org/10.1161/JAHA.120.016762.

[28] Chen Z, Ma X, Gao Y, Wu S, Xu N, Chen F, et al. Cardiac magnetic resonance-derived myocardial scar is associated with echocardiographic response and clinical prognosis of left bundle branch area pacing for cardiac resynchronization therapy. Europace: European Pacing, Arrhythmias, and Cardiac Electrophysiology: Journal of the Working Groups on Cardiac Pacing, Arrhythmias, and Cardiac Cellular Electrophysiology of the European Society of Cardiology 2023;25. Available from https://doi.org/10.1093/EUROPACE/EUAD326.

[29] Maisel AS, Krishnaswamy P, Nowak RM, McCord J, Hollander JE, Duc P, et al. Rapid measurement of B-type natriuretic peptide in the emergency diagnosis of heart failure. The New England Journal of Medicine 2002;347:161–7. Available from https://doi.org/10.1056/NEJMOA020233.

[30] Januzzi JL, Van Kimmenade R, Lainchbury J, Bayes-Genis A, Ordonez-Llanos J, Santalo-Bel M, et al. NT-proBNP testing for diagnosis and short-term prognosis in acute destabilized heart failure: an international pooled analysis of 1256 patients: the International Collaborative of NT-proBNP study. European Heart Journal 2006;27:330–7. Available from https://doi.org/10.1093/EURHEARTJ/EHI631.

[31] McMurray JJV, Packer M, Desai AS, Gong J, Lefkowitz MP, Rizkala AR, et al. Angiotensin–neprilysin inhibition versus enalapril in heart failure. New England Journal of Medicine 2014;5:132–3. Available from https://doi.org/10.1056/NEJMOA1409077.

[32] Biagini D, Pugliese NR, Vivaldi FM, Ghimenti S, Lenzi A, De Angelis F, et al. Breath analysis combined with cardiopulmonary exercise testing and echocardiography for monitoring heart failure patients: the AEOLUS protocol. Journal of Breath Research 2023;17:046006. Available from https://doi.org/10.1088/1752-7163/ACEC08.

[33] Gouzi F, Ayache D, Hédon C, Molinari N, Vicet A. Breath acetone concentration: too heterogeneous to constitute a diagnosis or prognosis biomarker in heart failure? A systematic review and meta-analysis. Journal of Breath Research 2021;16:016001. Available from https://doi.org/10.1088/1752-7163/AC356D.

[34] Schwarz K, Pizzini A, Arendacká B, Zerlauth K, Filipiak W, Schmid A, et al. Breath acetone-aspects of normal physiology related to age and gender as determined in a PTR-MS study. Journal of Breath Research 2009;3. Available from https://doi.org/10.1088/1752-7155/3/2/027003.

[35] Marcondes-Braga FG, Gioli-Pereira L, Bernardez-Pereira S, Batista GL, Mangini S, Issa VS, et al. Exhaled breath acetone for predicting cardiac and overall mortality in chronic heart failure patients. ESC Heart Failure 2020;7:1744–52. Available from https://doi.org/10.1002/EHF2.12736.

[36] Rydosz A. Diabetes without needles: non-invasive diagnostics and health management, diabetes without needles: non-invasive diagnostics and health management. 2022. p. 1–302. Available from https://doi.org/10.1016/B978-0-323-99887-1.01001-3.

[37] Marcondes-Braga FG, Gutz IGR, Batista GL, Saldiva PHN, Ayub-Ferreira SM, Issa VS, et al. Exhaled acetone as a new biomarker of heart failure severity. Chest 2012;142:457–66. Available from https://doi.org/10.1378/CHEST.11-2892.

CHAPTER

Medical data analysis support

8

Krzysztof Piotr Malinowski[1,2] **and Michał Szafarczyk**[1,2]
[1]*Center for Digital Medicine and Robotics, Jagiellonian University Medical College, Kraków, Poland,*
[2]*Functional and Virtual Medical 3D Imaging Laboratory, Department of Diagnostic Imaging, University Hospital in Kraków, Kraków, Poland*

8.1 Introduction

In the modern clinical landscape, the diagnostic process is no longer a solitary endeavor between a physician and a patient's symptoms. We have entered the era of the digital consultant, where computer-aided detection (CADe) and computer-aided diagnosis (CADx) serve as silent partners in the reading room. However, the bridge between a raw computational algorithm and a life-saving clinical decision is built not on code but on the rigorous discipline of biostatistics.

This chapter explores how biostatistical frameworks transform artificial intelligence (AI) from a "black box" into a transparent, reliable tool for medical decision support as well as the current state of the Software as a Medical Device (SaMD) landscape.

The applications of AI/machine learning (ML) within the SaMD framework span the entire clinical continuum—from triage and diagnosis to long-term chronic disease management. Because SaMD is a software intended for medical purposes that is not part of a hardware medical device, AI allows these "stand-alone" programs to analyze complex datasets (pixels, signals, or genomic sequences) to provide insights that were previously inaccessible [1].

8.2 Software as a medical device regulatory perspective

Clinical Decision Support (CDS) software encompasses a wide range of digital tools designed to aid healthcare professionals in making informed clinical decisions, but only some of these functions fall under FDA regulation. According to the FDA's guidance, updated in 2026, certain CDS functions are excluded from the definition of a medical device under Section 520(o)(1)(E) of the FD&C Act when they meet specific "Non-Device CDS" criteria, including enabling clinicians to independently review the basis of recommendations. While many decision-support

Functional and Virtual Medical 3D Imaging in Cardiovascular Medicine. DOI: https://doi.org/10.1016/B978-0-443-44570-5.00008-2

tools are non-device functions, others—particularly those that analyze medical images or signals, or those that clinicians cannot independently validate—remain regulated medical devices and are therefore subject to FDA oversight. The agency also stresses that the CDS guidance should not be applied in isolation; rather, it forms part of a broader ecosystem of digital health policies that together help determine the regulatory status of software functions [2].

Currently (December 5, 2025), there are 1357 FDA-cleared AI-enabled medical devices [3]. Radiology is the dominant panel, comprising more than three-quarters of the total, while several other specialties contribute only a small fraction each (Table 8.1). The list is, however, not exclusive as it contains only AI-enabled medical devices identified by scanning marketing authorizations and classifications for keywords defined in the FDA Digital Health and Artificial Intelligence Glossary [4].

The landscape of medical software has undergone a profound transformation, evolving from a niche technological curiosity into a cornerstone of modern clinical practice. Since 2016, the authorization of AI-enabled medical devices has accelerated at a staggering annual rate of approximately 49%, reflecting a broader industry shift toward the SaMD paradigm. This trajectory saw a dramatic increase from a mere six annual clearances in 2015 to a record 221 in 2023, with the cumulative total of authorized devices surpassing 1200 by mid-2025. This surge is

Table 8.1 List of lead panels with number of FDA-cleared artificial intelligence–enabled medical devices.

Panel (lead)	N	%
Radiology	1039	76.57%
Cardiovascular	130	9.58%
Neurology	61	4.50%
Anesthesiology	23	1.69%
Gastroenterology-urology	21	1.55%
Hematology	20	1.47%
Ophthalmic	10	0.74%
Clinical chemistry	9	0.66%
Pathology	8	0.59%
General and plastic surgery	6	0.44%
Microbiology	6	0.44%
Clinical toxicology	5	0.37%
Dental	5	0.37%
Orthopedic	5	0.37%
General hospital	4	0.29%
Obstetrics and gynecology	4	0.29%
Immunology	1	0.07%

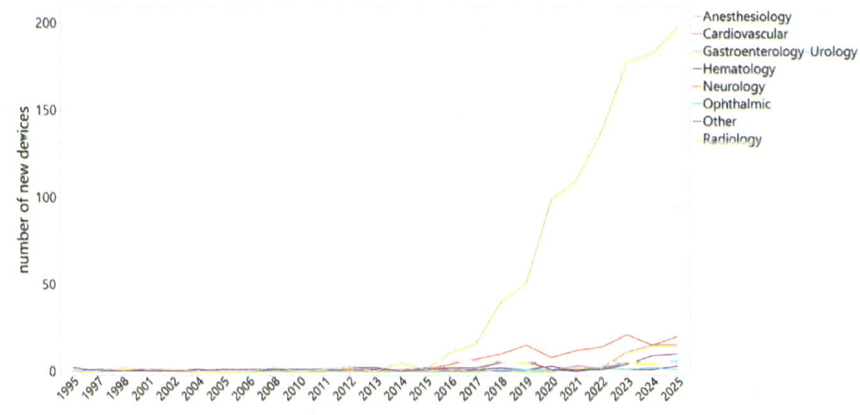

FIGURE 8.1

Number of new devices each year.

largely concentrated within the field of radiology, which consistently accounts for nearly 80% of all authorizations. The discipline's dominance is driven by the inherent compatibility of deep learning with high-volume imaging data, such as computed tomography (CT), magnetic resonance imaging (MRI), and X-ray scans. While other specialties are beginning to expand their footprint, they remain a distant second and third; cardiology represents roughly 10% of the market—primarily focused on ECG and arrhythmia analysis—while neurology accounts for approximately 3%, often through the development of specialized triage tools for acute conditions like stroke (Fig. 8.1).

8.3 Special case of continuous machine learning

A continual ML model is distinguished by its capacity to dynamically enhance and adjust its performance over time, integrating new data or experiences while simultaneously preserving previously acquired knowledge and information. This adaptability means that as the model is exposed to additional, updated datasets, it can refine its internal parameters and decision-making processes, resulting in outputs that may differ for the same set of inputs before and after the learning updates have been applied. The process of updating such models is meticulous and methodical, typically involving a series of steps such as data collection, preprocessing, retraining, validation, and rigorous performance analysis to ensure that the modifications lead to tangible improvements without compromising the integrity of past learning. These updates are not random or ad hoc; rather, they are governed by a clearly defined framework or protocol that prioritizes reliability, safety, and transparency, especially in critical domains like healthcare and medical software where accuracy and consistency are paramount. In stark contrast to locked models—which remain static and unresponsive to new information after their initial deployment—a continual learning system is built with mechanisms that allow ongoing evolution. This enables the model to remain relevant and effective in environments characterized by rapidly changing data patterns, shifting clinical guidelines, or emerging medical knowledge, thereby supporting sustained high-quality decision-making and patient care [5].

8.4 Data acquisition and preparation pipeline

The foundation of any medical data analysis project lies in the rigorous acquisition and preparation of the dataset. Unlike general computer vision, where images can often be scraped from the web, medical data acquisition is a strictly regulated process involving legal, ethical, and technical layers. The transition from a hospital Picture Archiving and Communication System to a clean, ML-ready dataset follows a complex pipeline requiring close collaboration between engineering teams and clinical partners.

8.4.1 Ethical compliance and 3D data anonymization

The process begins with obtaining approval from a Bioethical Committee (Institutional Review Board). Whether the study is retrospective (using historical data) or prospective, strict adherence to data protection regulations (such as GDPR in Europe or HIPAA in the United States) is mandatory. Once access is granted, the raw data—typically in Digital Imaging and Communications in Medicine (DICOM) format—must undergo a thorough de-identification process before leaving the hospital firewall. DICOM files contain extensive metadata tags (e.g., PatientName, PatientID, BirthDate) that must be removed or pseudonymized. While automated tools (e.g., DicomCleaner, RSNA DICOM Anonymizer) can efficiently scrub standard tags, "leakage" of Protected Health Information remains a significant risk in nonstandard acquisition protocols. A particular challenge arises when patient information is "burned in" to the pixel data itself—or for instance, when an ultrasound machine or an old CT scanner overlays the patient's name directly onto the image frames. In such cases, metadata scrubbing is ineffective. These datasets require advanced preprocessing steps, such as Optical Character Recognition to detect text within the image or manual cropping/masking of specific image regions to ensure complete anonymity without compromising the anatomical region of interest (ROI).

8.4.2 Ground truth generation: the segmentation protocol

Acquiring the image is only the first step; training supervised models requires high-quality labels, known as the ground truth. In 3D imaging, this typically involves volumetric segmentation—a labor-intensive process where anatomical structures are delineated slice by slice. Creating segmentation data is often the primary bottleneck in medical AI projects. It demands not only proficiency in annotation software (e.g., 3D Slicer, ITK-SNAP) but, more importantly, solid anatomical knowledge. To ensure consistency, especially when multiple annotators are involved, the research team must define a strict segmentation protocol. This document specifies exactly which structures are included in the label (e.g., include the renal arteries, but exclude branches) and how to handle ambiguous boundaries or artifacts. A poorly defined protocol leads to "label noise," which severely degrades model performance, regardless of the architecture used.

8.4.2.1 *Ground truth: bias and how to avoid it*

In the development of CADx systems, overfitting to the ground truth occurs when a model learns the specific subjective biases of the human observers rather than the objective biological reality of the disease. This problem becomes particularly acute when the same panel of experts provides the labels for the training, testing, and validation datasets. This creates a closed-loop validation environment that can lead to dangerously inflated performance metrics.

In the pivotal clinical trials for the SaMD used for regulatory purposes if the model is tested against the same panel that was used during algorithm development, it will appear highly accurate because it has successfully mimicked its expert panel used in the ground truthing of development datasets. However, if that model is deployed at a different hospital with a different panel of experts, its performance often collapses because it hasn't learned universal diagnostic features.

The issues of imperfect or biased ground truth can be mitigated during planning and statistical analysis. At the study design stage, choosing the strongest ground truth is crucial (Table 8.2).

To prevent this form of overfitting, biostatisticians and researchers use several strategies including having independent validation sets in pivotal study (the testing dataset should ideally be labeled by a completely different set of experts from a different institution who were not involved in the training phase and having hold-out "clean" data; reserve a portion of data labeled only by objective means [like

Table 8.2 Ground truth and risk of bias.

Truth type	Source	Risk of overfitting
Weak ground truth	Expert opinion/visual grading	High: Model learns human habits and biases.
Strong ground truth	Histopathology (biopsy)/ surgery	Low: Model learns biological reality.
Ultimate ground truth	Long-term clinical outcome (death/recovery)	Lowest: Model learns the actual impact on patient health.

biopsy] to serve as the final arbiter of performance, independent of the experts' visual interpretations, if possible).

Statistical techniques used to correct imperfect ground truth involve both frequentist and Bayesian approaches dealing with nonidentifiable models. When dealing with the estimation problem in a nonidentified statistical model—where the available data are insufficient to uniquely determine the values of all parameters—different methodological approaches are required to achieve meaningful inference. In the frequentist framework, this challenge is typically addressed by imposing additional constraints on the model parameters. These constraints might take the form of fixing certain parameters at specific values, enforcing equality or inequality relationships between parameters or introducing structural assumptions that reduce the space of possible solutions. By doing so, the model becomes identifiable, allowing for the estimation of parameters using the observed data.

In contrast, the Bayesian approach handles nonidentifiability by incorporating prior information through the specification of a proper prior distribution over the parameters. This prior reflects existing knowledge or plausible values for the parameters based on previous studies, expert opinion, or theoretical considerations. Bayesian inference then combines this prior distribution with the likelihood derived from the observed data to produce the posterior distribution, which forms the basis for making inferences about the parameters. The informativeness of the prior is especially crucial in nonidentified models, because the data alone do not provide sufficient guidance. As a result, to obtain useful and interpretable parameter estimates, the Bayesian method requires that the prior distribution be informative for at least as many parameters as would be fixed or otherwise constrained in the frequentist approach. This ensures that the posterior distribution is well-defined and the resulting inferences are meaningful, leveraging both prior knowledge and observed evidence to compensate for the model's inherent limitations [6].

Another important source of bias can arise in diagnostic test studies when not all patients who undergo the test have their condition status confirmed by a definitive ground truth–generation procedure. This situation is referred to as verification bias or workup bias. Unlike random sampling, the subgroup of patients who do not receive ground truth verification is typically selected based on certain characteristics, which can systematically influence the study's outcomes. For instance, consider a clinical scenario where the ground truth is established through an invasive procedure such as surgery or biopsy. In these cases, patients who receive negative results from the initial, noninvasive diagnostic test are often less likely to proceed to surgery, since the risks and costs associated with the procedure may not be justified for them. Conversely, those with positive test results are much more likely to undergo the confirmatory, invasive evaluation, as clinicians seek to verify and treat suspected disease. While this selective verification process is often practical and economically sensible in routine clinical care, it introduces a significant methodological concern in research settings designed to assess the accuracy of diagnostic tests. Because the group of patients verified by the ground truth is enriched with individuals more likely to have the condition, the sensitivity and specificity calculated from these data may not accurately reflect the true performance of the test in the broader patient population. In particular, the apparent accuracy may be inflated, as false negatives and true negatives are underrepresented among those who are verified. This can lead to an overestimation of the test's diagnostic capability, potentially resulting in misguided clinical decision-making if the bias is not properly accounted for during analysis. Researchers must recognize and address verification bias through careful study design, statistical adjustment, or alternative validation strategies to ensure that the reported accuracy measures are both reliable and applicable to real-world clinical practice. This issue can be easily overcome by applying the missing data framework [7].

8.4.3 The role of open access data

While proprietary hospital datasets offer the advantage of specific clinical context, the democratization of medical AI has been significantly accelerated by open access repositories. These public datasets serve as essential benchmarks for comparing model architectures. Notable examples include the datasets released alongside challenges organized by the Radiological Society of North America [8], which provide high-quality, expert-annotated CT and MRI scans for tasks like hemorrhage detection or fracture localization. Beyond imaging, repositories like PhysioNet host massive signal datasets, such as PTB-XL [9], which contains approximately 22,000 ECG records from nearly 19,000 patients. These large-scale, heterogeneous collections are invaluable for pretraining models and validating their generalization capabilities across different populations before fine-tuning them on smaller, institution-specific datasets (Fig. 8.2).

8.4.3.1 "Name The Dataset!" bias in open access data

The "Name That Dataset!" bias refers to a phenomenon where AI models learn to identify the specific source or origin of a medical image, such as the hospital or the specific machine that produced it, rather than the actual clinical pathology. In open access environments, datasets like NIH [10], CheXpert [11], MIMIC-CXR [12], and PadChest [13] can contain fingerprints including unique machine artifacts, varying image resolutions, institution-specific text overlays, specific semantic maps, pixel intensity, or texture characteristics. When models are trained across these diverse but distinct repositories, they frequently take those fingerprints as features used in diagnosis or prediction that can inflate the model performance, as those fingerprints will not be available in the real-world scenario.

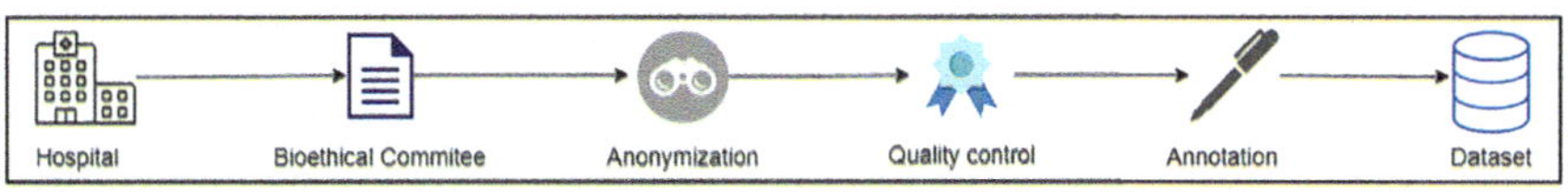

FIGURE 8.2

The medical data acquisition pipeline.

To check whether the bias exist in particular setting, a model can be fitted to predict the source or origin of the dataset and to identify aforementioned fingerprints. Such models, particularly convolutional neural networks (CNNs), consistently attained high accuracy in standard dataset classification tasks [14].

There are, however, several ways to counter the bias while working on open access data. The gold standard for proving a model hasn't overfitted to a specific dataset is external validation. Biostatisticians mandate that the software/algorithm be tested on data from institutions that contributed zero images to the training set. The best approach for that dataset is to prospectively collect it after the training process is completed and model development has stopped. If the validation on such a dataset revealed that the model overfits the data, it can be included in the development process, and a new validation dataset must be prospectively collected. Such an approach, however the best, can drastically inflate the cost of model training and certification as SaMD. Other approaches can be used during the training process to minimize the number of validation datasets used (although at least one prospective dataset must be used in almost all SaMDs). One of such solutions is adversarial training in which researchers can train a secondary network that tries to guess the data's origin and the primary diagnostic model is then penalized if that secondary network succeeds. This forces the algorithm to ignore dataset-specific fingerprints and focus only on the biological signals that are universal across all data sources (institutions, hospital, machines, etc.).

8.5 Artificial intelligence/machine learning in 3D imaging

8.5.1 Overview

The application of ML in medicine dates back to the 20th century. However, early solutions focused primarily on tabular data, utilizing Bayesian inference and expert-designed features/handcrafted features [15]. The turn of the 21st century marked a paradigm shift towards neural networks. Although the concept of CNNs emerged in the 1980s and 1990s, computational limitations delayed their widespread adoption.

A pivotal moment occurred in 2009 with the release of the ImageNet dataset, which catalyzed the rapid development of computer vision architectures. This led to the introduction of AlexNet in 2012, followed by residual networks (ResNets) and U-Net in 2015. For the medical domain, U-Net constituted a breakthrough innovation. Designed specifically for biomedical image segmentation, its encoder-decoder architecture with skip connections remains a foundational standard in the field to this day.

8.5.2 Core applications in medical imaging

The utility of ML in medical imaging can be categorized into several primary objectives. Generally, the goal is to develop a decision support system that acts as an expert consultant, providing clinicians with quantitative insights during diagnosis or patient care.

1. Classification: The primary purpose of classification models is to function as reliable predictors for specific pathologies based on imaging data. For instance, a model may process a patient's head CT scan as input and output a probability score indicating the likelihood of the presence of a brain aneurysm.
2. Diagnostic segmentation (pathology localization): Unlike simple classification, diagnostic segmentation tools must not only identify the condition but also precisely localize it within the volumetric data. A common application involves oncology, such as breast cancer diagnosis. In this scenario, the model processes the medical image and outputs a probability score for each voxel (in 3D) or pixel (in 2D). This results in a heat map overlaid on the original scan, where regions suspected of being tumoral exhibit higher intensity values. Consequently, the physician receives an augmented visualization highlighting suspicious areas, which can then be verified or discarded as false positives.
3. Segmentation for anatomical modeling: While similar in method to diagnostic segmentation, this category differs fundamentally in the utility of its output. In the previous section, models served as decision support for diagnosis. Conversely, segmentation for modeling is often an intermediate step in a larger workflow rather than the final product. Segmented structures are essential for downstream applications, including visualization in virtual reality, mixed reality, and augmented reality (VR/MR/AR) or for creating physical models via 3D printing. These models automate the generation of accurate anatomical representations, a process that—when performed manually—is labor-intensive and requires significant anatomical expertise. Automated segmentation can save hours of manual contouring; however, these models face limitations, particularly when encountering congenital or acquired anatomical anomalies, making their performance heavily dependent on the diversity and quality of the training data.

8.5.3 Key model architectures in medical imaging

The transition from manual feature engineering to deep learning has been driven by the development of specialized neural network architectures. While generic computer vision models provide a strong starting point, the specific requirements of medical imaging—such as the need for volumetric processing and high precision with limited data—have necessitated the adaptation of these architectures. Currently, three families of models dominate the landscape: ResNets, U-Net–based architectures, and the emerging vision transformers (ViTs) (Fig. 8.3).

8.5.3.1 Residual networks: solving the depth problem

CNNs extract features hierarchically: shallow layers detect edges and textures, while deeper layers identify complex anatomical structures. However, historically, training very deep networks proved difficult due to the "vanishing gradient"

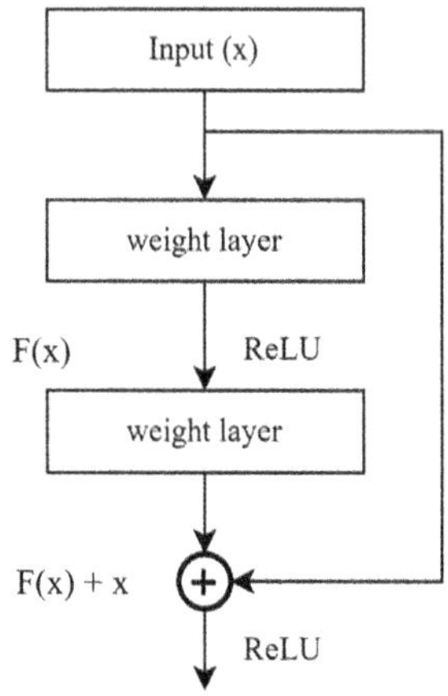

FIGURE 8.3

The residual block (ResNet architecture). Schematic representation of a residual building block. The skip connection (curved arrow) allows the gradient to bypass the nonlinear transformations, mitigating the vanishing gradient problem and enabling the training of very deep networks. Image redrawn from He, K, Zhang, X, Ren, S, Sun, J. Deep residual learning for image recognition, In: Proceedings of the IEEE conference on computer vision and pattern recognition; 2016;770–8 [16].

problem, where the signal required to update weights diminishes as it back-propagates through layers. ResNet revolutionized this by introducing skip connections (or residual connections). Instead of learning a direct mapping, the network learns a residual function. Formally, a building block is defined as:

$$y = F(x, \{W_i\}) + x$$

where x is the input and y is the output. This simple addition allows gradients to flow through the network unimpeded. In medical imaging, ResNet variants (e.g., ResNet-50, ResNet-101) serve as the standard "backbone" for classification tasks (e.g., distinguishing benign from malignant lesions) and as feature extractors for more complex segmentation models (Fig. 8.4).

8.5.3.2 U-Net: the gold standard for segmentation

While ResNet excels at classification, medical diagnosis frequently requires segmentation—pixel-wise or voxel-wise delineation of structures. The U-Net architecture was designed specifically for this purpose and remains the most influential architecture in medical image analysis. The architecture consists of two symmetric paths, giving it a "U" shape:

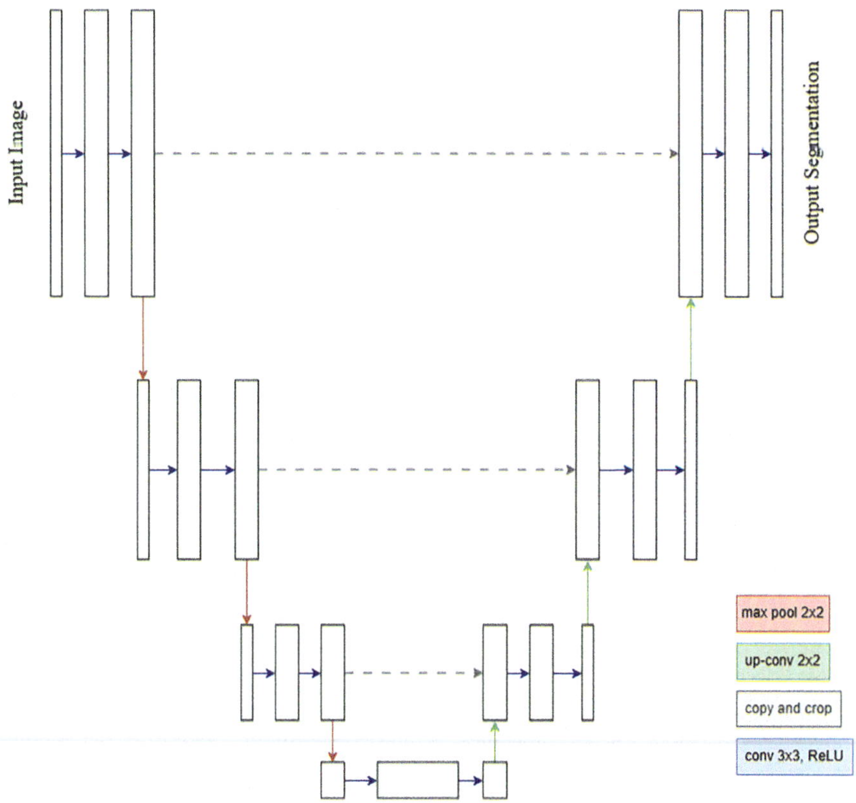

FIGURE 8.4

The U-Net architecture. The U-Net architecture designed for biomedical image segmentation. It consists of a contracting path (left) to capture context and a symmetric expanding path (right) that enables precise localization. The horizontal arrows represent skip connections that transfer high-resolution feature maps from the encoder to the decoder.

Image redrawn from Ronneberger O., Fischer P., Brox T. U-net: convolutional networks for biomedical image segmentation. In: International conference on medical image computing and computer-assisted intervention. Cham: Springer international publishing; 2015;234–41 [17].

- The contracting path (encoder): Captures context and reduces spatial resolution (similar to a standard CNN).
- The expansive path (decoder): Upsamples the feature map to restore the original resolution for precise localization. Crucially, U-Net employs skip connections that concatenate high-resolution features from the encoder directly to the decoder. This allows the model to combine semantic information (what is the object?) with spatial information (where is the object?). For 3D imaging, the 3D

U-Net and V-Net extend this concept to volumetric data using 3D convolutions. Recently, the nnU-Net (no-new-U-Net) [18] framework has emerged as the state-of-the-art, automatically configuring the U-Net architecture (preprocessing, topology, and training strategy) based on the specific dataset properties, demonstrating that a well-tuned U-Net is often superior to more complex architectural modifications.

8.5.3.3 You Only Look Once

While U-Net excels at delineating precise boundaries, certain clinical scenarios prioritize speed and localization over pixel-perfect morphology. This domain is addressed by Object Detection models, with the You Only Look Once (YOLO) [19] family representing the gold standard for real-time inference. Unlike traditional two-stage detectors (such as R-CNN) that first generate region proposals and then classify them—a computationally expensive process—YOLO reframes object detection as a single regression problem.

The architecture divides the input image into an $S \times S$ grid. Each grid cell is responsible for predicting B bounding boxes and confidence scores for objects whose center falls within the cell. This "single-shot" approach allows the entire image to be processed in one forward pass through the neural network, achieving inference speeds capable of processing video streams at high frame rates (often exceeding 60 FPS).

In medical imaging, this capability is critical for intraoperative assistance and video-based diagnostics. For instance, in gastroenterology, YOLO-based models are deployed during live colonoscopies to flag potential polyps for the endoscopist in real-time, reducing the "miss rate" due to fatigue. Similarly, in ultrasonography, YOLO is used to automatically localize anatomical landmarks or guide the probe plane. Furthermore, YOLO often serves as an efficient preprocessing step: it detects and crops the ROI containing a pathology, which is then passed to a heavier segmentation model (like U-Net) for detailed analysis, thereby optimizing the overall computational pipeline.

8.5.3.4 Vision transformers: capturing global context

Convolutional networks have an inherent limitation: they possess a local receptive field. A convolution operation only "sees" a small neighborhood of pixels at a time, making it difficult to model long-range dependencies—or for instance, understanding the spatial relationship between widely separated organs in a full-body CT scan. Inspired by their success in natural language processing, transformers utilize a self-attention mechanism. ViTs [20] treat an image as a sequence of patches (analogous to words in a sentence). This allows the model to weigh the importance of every part of the image relative to every other part, regardless of the distance. While ViTs require larger datasets to train effectively compared to CNNs, they are increasingly used in hybrid architectures (e.g., TransUNet [21] or UNETR [22]). These hybrids combine the local feature extraction efficiency of CNNs with the global context understanding of transformers, offering superior performance in complex segmentation tasks where understanding the overall anatomical context is critical.

8.5.4 2D, 2.5D, and 3D architectures

Medical imaging modalities such as CT and MRI are inherently volumetric. A single examination consists of a stack of 2D slices forming a 3D voxel grid. Choosing the appropriate dimensional strategy for processing this data (2D, 2.5D, or fully 3D) is one of the most critical design decisions in medical deep learning, involving a trade-off between computational efficiency, data availability, and spatial context awareness.

8.5.4.1 *2D architectures (slice-based)*

In this approach, the 3D volume is decomposed into individual 2D slices (e.g., along the axial plane), which are processed independently by the network.

This method is computationally efficient and has lower hardware requirements, allowing for larger batch sizes. Crucially, it enables the use of models pretrained on massive natural image datasets (like ImageNet) via transfer learning, which significantly accelerates convergence when medical data is scarce.

The primary drawback is the loss of volumetric context. Since the model treats each slice in isolation, it ignores the anatomical continuity along the z-axis (inter-slice correlation). This often leads to inconsistent predictions, where a structure detected in slice N disappears in slice $N + 1$, resulting in "jittery" or discontinuous 3D segmentations.

8.5.4.2 *3D architectures (volumetric)*

Full 3D architectures (e.g., 3D U-Net, V-Net) replace 2D convolutions with 3D kernels (e.g., $3 \times 3 \times 3$), processing the entire volume or large sub-volumes (patches) simultaneously.

Theoretically, these models capture the most complete spatial context. The "curse of dimensionality" results in a cubic increase in parameters and memory usage. Training 3D networks requires significant GPU VRAM, frequently forcing small batch sizes (which destabilizes batch normalization). Furthermore, medical data is frequently anisotropic (high resolution in X-Y plane, low in z-axis). 3D convolutions typically struggle to learn effective features from such inconsistent physical dimensions, leading to suboptimal performance compared to lighter models [18].

8.5.4.3 *2.5D architectures (pseudo-3D)*

2.5D approaches seek an optimal trade-off. Instead of a single slice, the network receives a stack of adjacent slices (e.g., $N - 2$ to $N + 2$) as input channels to predict the label for the central slice.

This strategy is widely regarded as the "sweet spot" in clinical deployment. It incorporates sufficient volumetric context to ensure continuity while maintaining the computational efficiency of 2D networks. Recent benchmarks demonstrate that for many tasks (e.g., liver tumor segmentation), 2.5D models achieve accuracy comparable to or exceeding full 3D models, particularly when training data is limited [23].

Contrary to early assumptions, 3D architectures are not inherently superior. Current evidence suggests that for anisotropic data (e.g., standard clinical CTs with thick slices), 2D or 2.5D ensembles often outperform full 3D models due to easier optimization and better use of pretrained weights. The choice must therefore be data-driven: utilize 3D only when isotropic high-resolution data is available and spatial continuity is the dominant diagnostic feature.

8.5.5 Feature extraction and integration of tabular data

While deep learning (DL) has revolutionized the analysis of visual patterns, medical decision-making is rarely based on imaging alone. A comprehensive diagnosis relies on a multimodal approach, integrating visual data with clinical context stored in electronic health records (EHR), such as demographics, laboratory results, and genomic profiles. This section explores two critical pathways: the extraction of quantitative features from images (Radiomics) and the fusion of this information with tabular clinical data.

Radiomics operates on the hypothesis that medical images contain quantifiable data reflecting underlying pathophysiology that is imperceptible to the human eye. Unlike deep learning, where features are learned implicitly, radiomics involves the mathematical extraction of explicit, handcrafted features from a segmented ROI according to the Image Biomarker Standardization Initiative. This process generates a high-dimensional feature space describing the tumor phenotype. Basic descriptors include first-order statistics, such as intensity distribution (mean, skewness, kurtosis). More complex, shape-based features quantify geometric properties like sphericity or surface-area-to-volume ratio. Finally, texture analysis—utilizing matrices such as the Gray Level Co-occurrence Matrix—captures the spatial interrelationships between voxels, serving as a potent proxy for tissue heterogeneity. These "radiomic signatures" are subsequently processed using classical ML algorithms like random forests or XGBoost to predict clinical endpoints, such as immunotherapy response.

Integrating EHR data presents unique challenges distinct from imaging. Clinical datasets are inherently heterogeneous, mixing continuous variables (e.g., leukocyte count) with categorical data (e.g., tumor stage) and unstructured text. Furthermore, real-world data is notoriously incomplete, requiring robust imputation strategies ranging from simple statistical substitution to multivariate methods like MICE. A critical aspect of medical tabular data is class imbalance, where pathological cases are significantly outnumbered by controls. While synthetic oversampling techniques like SMOTE have been historically utilized, their application in medicine is controversial. Linear interpolation between patients often yields biologically implausible samples that introduce noise rather than signal. Consequently, current best practices favor cost-sensitive learning—assigning higher penalties to misclassifying minority cases—over balanced sampling strategies that optimize the training process without artificially synthesizing patient data.

The convergence of imaging and clinical data requires specific architectural choices to effectively combine these distinct modalities. Three primary fusion strategies have emerged in the literature:

- Early fusion (input-level): Clinical variables are concatenated with the image data at the input stage. For example, demographic data can be appended as additional channels to the image. However, this is often suboptimal due to the dimensional disparity between a sparse vector and a dense pixel grid.
- Late fusion (decision-level): Separate models are trained for each modality (e.g., a CNN for the image and an XGBoost model for tabular data). Their final predictions (probabilities) are then aggregated, often using a weighted average or a meta-learner (stacking). This approach is robust and allows for handling missing modalities flexibility.
- Joint (intermediate) fusion: This involves extracting a feature vector from the deep neural network (e.g., the bottleneck layer of a ResNet) and concatenating it with the processed tabular data. This combined vector is then passed through a final set of fully connected layers to make the prediction. This allows the model to learn nonlinear interactions between the image features and clinical variables (e.g., "this texture pattern is only dangerous if the patient is over 60").

8.5.6 Data augmentation for medical data

Deep learning models are notoriously data-hungry, yet medical datasets are frequently limited in size due to privacy concerns and the high cost of expert annotation. Data augmentation serves as a critical regularization technique, artificially expanding the training set diversity to prevent overfitting. However, unlike in natural image processing, augmentations in medicine must remain physically and anatomically plausible. An aggressive transformation that is valid for general computer vision might inadvertently simulate a nonexistent pathology or anatomically impossible configuration in a medical scan.

The most fundamental augmentations involve geometric manipulations designed to introduce invariance to position and orientation. These include rigid transformations such as rotation, translation, and scaling. In the context of soft-tissue analysis—or for instance, in MRI of the liver or stomach—elastic deformations are particularly valuable. By applying random smooth deformations, the network learns to handle the natural variability of organ shapes and the pressure exerted by surrounding tissues, a technique that was instrumental in the success of the original U-Net architecture. However, care must be taken with operations like mirroring;

while flipping a brain MRI is generally valid, flipping a chest X-ray could confuse the model regarding heart laterality or conditions like situs inversus.

Beyond geometry, effective augmentation must simulate the physics of image acquisition. Medical images are physical measurements, and models must be robust to hardware variability. Techniques in this category include the injection of Gaussian or Rician noise to simulate varying signal-to-noise ratios found across different scanner generations. Similarly, random intensity shifting and scaling mimic the differences in contrast protocols (e.g., window levels in CT). In MRI analysis, simulating bias fields—low-frequency intensity variations caused by magnetic-field inhomogeneities—forces the model to rely on structural features rather than absolute intensity values, significantly improving generalization across different clinical centers.

A rapidly evolving frontier is the use of generative adversarial networks [24] and denoising diffusion probabilistic models [25] to generate entirely new, synthetic medical images.

- Cross-modality synthesis: Using models like CycleGAN to transform images from one domain to another (e.g., synthesizing a CT scan from an MRI). This is valuable when one modality is scarce (e.g., creating synthetic CTs for radiotherapy planning based on available MRIs).
- Pathology inpainting: Generative models can be trained to "paint" tumors into images of healthy patients, creating diverse, labeled training examples for rare pathologies. However, rigorous validation is required to ensure these synthetic data remain biologically realistic.

8.5.7 Virtual human twin

The concept of the virtual human twin (VHT) represents the apex of medical data analysis support. While traditional diagnostic tools provide a snapshot of the patient's current state, a VHT is a dynamic, multiscale representation designed to simulate physiological processes and predict future trajectories. It signifies a paradigm shift from reactive medicine to predictive, personalized healthcare. This transition is powered by the convergence of the technologies discussed in previous sections: high-fidelity 3D imaging provides the anatomical geometry, while ML on tabular and genomic data informs the physiological boundary conditions.

Constructing a VHT begins with the automated processing of medical images. The AI-driven segmentation models (such as U-Net or nnU-Net) transform raw volumetric data (CT/MRI) into precise 3D surface meshes. However, geometry alone is insufficient. To create a functional twin, these anatomical models are integrated with mechanistic solvers—computational frameworks based on the laws of physics, such as computational fluid dynamics for blood flow or finite element analysis for tissue mechanics. Historically, these simulations were computationally prohibitive, often requiring weeks on supercomputers to model a single cardiac cycle. Today, deep learning serves as a bridge: AI-based "surrogate models" and physics-informed neural networks can approximate these complex physical simulations in near real-time, making patient-specific modeling feasible within a clinical timeframe.

The realization of this technology requires an ecosystem that transcends individual research labs. In Europe, this effort is spearheaded by the EDITH (Ecosystem for Digital Twins in Healthcare) project, a coordination action funded by the European Commission. EDITH aims to map the roadmap for a fully integrated VHT, addressing the fragmentation of current solutions. The initiative focuses on creating a federated repository of models and data, ensuring that a "Heart Twin" developed in one center can communicate with a "metabolic twin" from another. This interoperability is crucial for the ultimate goal: moving away from generic treatment guidelines toward in silico trials, where therapies are tested on a patient's digital replica before being administered in reality.

The immediate application of VHT lies in surgical planning and intervention. For instance, in cardiovascular medicine, digital twins are already used to calculate fractional flow reserve noninvasively from CT scans, eliminating the need for invasive catheterization. In oncology, "digital twins" of tumors are being developed to predict drug response based on the specific vascularization and genetic makeup of the lesion. Key pillars for the future development of VHT include:

- Multiscale integration: Linking processes from the molecular level (genomics) to the organ and systemic level (whole-body imaging).
- Continuous updates: The twin is not static; it evolves. Data from wearable devices (smartwatches, biosensors) must be continuously fed into the model to update the twin's state in real-time.
- Ethical and legal frameworks: Determining liability when a digital twin's prediction fails remains a significant challenge that initiatives like the European Health Data Space (EHDS) aim to address.

8.6 Model validation: data science perspective

8.6.1 Metrics for validation

Validation in medical image analysis is not merely a measure of algorithmic performance but a critical assessment of safety and clinical utility. Unlike general computer vision tasks, where a misclassification might be trivial, errors in medical diagnosis can have life-altering consequences. Therefore, the validation framework must move beyond simple accuracy, which is often misleading in highly imbalanced medical datasets—towards nuanced metrics that capture sensitivity, spatial overlap, and boundary precision. Furthermore, the choice of the loss function during training acts as a fundamental driver for the model's behavior, dictating how errors are penalized and how the network prioritizes different aspects of the learning task.

The evaluation of medical models depends heavily on the specific clinical task. For classification tasks, such as detecting the presence of a pathology, the primary concern is the balance between identifying true cases and avoiding false alarms. This is quantified by sensitivity (recall) and specificity. Sensitivity measures the proportion of actual positives correctly identified, while specificity measures the proportion of negatives correctly identified.

$$Sensitivity = \frac{TP}{TP + FN}, \quad Specificity = \frac{TN}{TN + FP}$$

In clinical practice, a high sensitivity is typically prioritized for screening tools to minimize missed diagnoses (false negatives), even at the cost of lower specificity. To evaluate the model's discriminative ability across various decision thresholds, the receiver operating characteristic (ROC) curve and the area under the curve (AUC) are standard metrics, providing a threshold-independent performance summary.

For segmentation tasks, where the goal is to delineate anatomical structures, pixel-wise accuracy is insufficient because the background often dominates the image. The industry standard for evaluating spatial overlap is the dice similarity coefficient (DSC), also known as the F1 score. The DSC quantifies the harmonic mean of precision and sensitivity, ranging from 0 (no overlap) to 1 (perfect overlap):

$$DSC = \frac{2\,|X \cap Y|}{|X| + |Y|} = \frac{2TP}{2TP + FP + FN}$$

where Y represents the ground truth and X represents the model prediction. While DSC is excellent for measuring overall volume agreement, it is relatively insensitive to boundary errors in large structures. A model might achieve a high DSC while missing a clinically critical spur or margin. To address this, geometric metrics such as the Hausdorff distance (HD) are employed. Specifically, the 95th percentile HD measures the maximum distance between the surface of the predicted segmentation and the ground truth, effectively penalizing the "worst-case" contouring error. This metric is particularly vital in fields like radiotherapy planning, where boundary precision dictates the safety margin for radiation delivery.

8.6.2 Loss functions

The loss function serves as the mathematical objective that the neural network attempts to minimize during training. The choice of loss function must align with the distribution of the data and the desired evaluation metric. The standard cross-entropy loss, while effective for classification, often struggles in segmentation tasks involving small ROIs due to the overwhelming dominance of background pixels. This class imbalance causes the gradient from the background to suppress the signal from the target structure, leading to models that default to predicting "background" everywhere.

To counteract this, the Dice loss [26] was introduced to optimize the Dice coefficient directly. By formulating the loss as *1 - DSC*, the network is driven to maximize the overlap regardless of the background size.

$$L_{Dice} - 1 - \frac{2\sum_i y_i \hat{y}_i}{\sum_i y_i + \sum_i \hat{y}_i}$$

However, Dice loss can be unstable during training when the intersection is small. A more robust alternative for handling difficult cases is focal loss [27] $(1 - p_t)^\gamma$. Originally designed for dense object detection, focal loss reshapes the standard cross-entropy loss by adding a modulating factor $(1 - p_t)^\gamma$ that reduces the loss contribution from easy, well-classified examples.

$$L_{Focal} = -(1 - p_t)^\gamma \log(p_t)$$

This forces the model to focus its learning capacity on "hard" examples—such as small lesions or unclear boundaries—that are frequently misclassified, rather than accumulating negligible gradients from the vast, easily classified background. Modern approaches often employ a combo loss, summing Dice and cross-entropy/focal loss, to benefit from both the smooth convergence of distribution-based losses and the overlap optimization of region-based losses.

8.6.3 Explainability and model transparency

The adoption of deep learning in clinical workflows faces a significant psychological and regulatory barrier: the "black box" phenomenon. While neural networks often outperform human experts in specific tasks, their internal decision-making processes remain opaque. In high-stakes medical environments, a prediction—no matter how accurate—is of limited value if the clinician cannot understand the rationale behind it. Consequently, explainable AI has transitioned from an academic curiosity to a regulatory necessity, mandated by frameworks such as the European AI Act. To bridge the gap between algorithmic probability and clinical reasoning, validation must extend beyond performance metrics to include interpretability and uncertainty estimation.

8.6.3.1 *Visualizing model attention*

For imaging models, interpretability is primarily achieved through post-hoc visualization techniques that highlight the regions influencing the model's decision. Saliency maps, and specifically gradient-weighted class activation mapping (Grad-CAM) [28], have become the industry standard for CNNs.

Grad-CAM utilizes the gradients flowing into the final convolutional layer to produce a coarse localization map, highlighting the important regions in the image. For instance, in a model trained to detect pneumonia, Grad-CAM should highlight the lung opacity. If the model instead highlights a metal implant or a text marker in the corner of the X-ray, the clinician immediately knows the prediction is based on a confounding artifact rather than pathology. This visual verification allows for the detection of "Clever Hans" behaviors, where models learn spurious correlations rather than genuine causal features.

8.6.3.2 *Feature attribution and uncertainty*

When dealing with multimodal or tabular data, the challenge shifts to quantifying the contribution of specific clinical variables. Techniques based on cooperative game theory, such as SHapley Additive exPlanations (SHAP) [29], provide a unified measure of feature importance. SHAP values explain a specific prediction by calculating the contribution of each feature (e.g., age, gene expression, tumor volume) to the difference between the actual prediction and the average baseline.

However, knowing *why* a model made a decision is insufficient; clinicians must also know *how confident* the model is. Standard softmax probabilities are often calibrated poorly and can be overconfident even on erroneous predictions. To address this, uncertainty estimation techniques like Monte Carlo dropout are employed. By keeping dropout layers active during inference and running multiple forward passes for the same input, the system generates a distribution of predictions rather than a single point estimate. The variance of this distribution serves as a proxy for uncertainty. If the variance is high, it signals that the input data (e.g., a rare anomaly) lies outside the model's training distribution, alerting the physician that human review is mandatory.

8.7 Model validation: biostatistics perspective

Similarly to the data science perspective, both sensitivity and specificity play a crucial role in validation of algorithm performance. However, measures like F1 score or Dice coefficient, that do not have clear clinical interpretations, are not usually used by biostatisticians, as they do not translate clarity to the clinical decision-making process. To counter this, biostatisticians developed a range of tools that can provide more patient-oriented validation, but before those, the overall approach for model validation must be described.

The first step is to define a validation endpoint, keeping in mind that if the algorithm must be useful, it must provide information that can be included in the medical decision-making process that often involves the entire human being. The endpoint should typically reflect the performance on a patient level, rather than an image or pixel level. For example, it is more important to determine if the patient has a tumor than to determine if the tumor is present on the image. If multiple images per patient are analyzed, the true positive (TP) should be declared on a patient level if at least "n" images contained tumor, and such TP should be used in estimation of sensitivity with denominator being all patients with tumor per gold standard (e.g., biopsy). Image-level analysis can also be informative and should be presented (TP would be an image with a detected tumor, and the denominator would be all images with tumor per ground truth) but usually not as a primary performance metric.

Similarly to image analyses, it is more significant to determine if a patient has arterial fibrillation than to determine if the specific ECG sample at a specific lead belongs to AF, even if the algorithm is trained to localize the AF in the signal. Detailed localization can be a secondary endpoint, but the primary endpoint associated with clinical decision-making process is frequently associated with a patient. However, in case several measurements per patient are obtained, for example, 2 CT scans or 2 ECGs, each CT/ECG can be treated separately, with appropriate adjustment for correlated observations.

Other endpoints can reflect the algorithm's predictive values, including positive predictive value (PPV) and negative predictive value (NPV). Both metrics, however, are correct only if the prevalence of the disease in question in the validation sample is the same as in target population to which the algorithm will be applied, which is usually not the case; hence, the PPV and NPV mostly used in model validation are prevalence adjusted, calculated from the Bayes theorem:

- PPV is defined as $PPV = \frac{Se * \gamma}{Se * \gamma + (1 - \mathrm{Sp})(1 - \gamma)}$
- NPV is defined as $NPV = \frac{Sp(1 - \gamma)}{Sp(1 - \gamma) + (1 - \mathrm{Se}) * \gamma}$

Where γ is the prevalence of the disease in question in the target population.

Other endpoints, especially those used in the case of image analyses, are associated with the localization ROC (LROC) and free-response ROC framework (FROC) as an extension of the ROC approach mentioned before.

For diagnostic tests designed to locate and assess tumors, estimators of sensitivity must require that the tumor be correctly located. If a tumor is found but the location does not match the gold standard's location of the tumor, then the detected tumor is a false positive and the missed true lesion is a false negative. A second important consideration in defining the accuracy of these diagnostic and screening tests is the number of true tumors among patients with the condition. If there is usually one tumor among patients with the condition, then patient-level estimators of diagnostic accuracy are appropriate. If diseased patients often have more than one lesion, then lesion-level estimators of diagnostic accuracy are needed as well, depending on the possible clinical decisions that can be made based on the algorithm prediction.

Another crucial aspect in defining the accuracy of these diagnostic and screening algorithms is considering the number of true tumors present among patients who have the condition. In cases where there is typically only one tumor per patient, patient-level estimators of diagnostic accuracy, such as sensitivity and specificity calculated on a per-patient basis, are generally sufficient and most clinically relevant. However, in many diseases, patients may present with multiple tumors. In these situations, tumor-level estimators of diagnostic accuracy become necessary as well. Tumor-level analysis evaluates how effectively the algorithm detects and localizes each individual tumor, which can be important for guiding clinical decisions such as determining the extent of the disease, planning multifocal interventions, or monitoring disease progression.

The choice between patient-level and tumor-level validation metrics should be guided by the clinical context and the types of decisions that will be informed by the algorithm's predictions. For instance, if treatment decisions depend on identifying every tumor (e.g., in metastatic cancer), tumor-level sensitivity and specificity offer additional insight beyond patient-level metrics. In summary, robust validation of diagnostic algorithms in radiology and related fields requires careful consideration of both the accuracy of detection and the precision of localization, with performance metrics tailored to the clinical realities and needs of the patient population.

The LROC approach is a specialized method for evaluating diagnostic algorithms, particularly in the context of medical imaging where the goal is to detect the presence and location of tumors. In this framework, the algorithm first analyzes the image and assigns a rating score that reflects the probability of there being at least one tumor anywhere within the image. Based on this score, the algorithm then identifies and highlights the single region it considers most suspicious for having a tumor. Unlike other approaches that might recognize multiple lesions, the LROC method is designed to only credit the algorithm for the most prominent finding, regardless of whether multiple tumors are actually present. This means that if there are several tumors in the image, the algorithm only receives recognition for correctly identifying the most suspicious one and does not gain additional credit for detecting additional lesions. As such, the LROC approach is most appropriate in clinical scenarios where patients typically have no more than one tumor, such as in certain screening contexts or diseases where multifocality is rare. This focus ensures that performance metrics remain clinically relevant and straightforward, aligning the evaluation with decision-making processes that prioritize the detection of a single, actionable finding per patient [30].

In the FROC method, the diagnostic algorithm is tasked with identifying and localizing all regions within an image that it considers suspicious for containing tumors. For each suspected tumor, the algorithm assigns a confidence score, which can be based on a standardized scale—such as a 5-, 7-, or 9-point Likert scale or a continuous percentage probability (e.g., 10%, 20%, etc.). This approach allows the assessment of both the detection and localization capabilities of the algorithm, as it must not only recognize potential tumors but also pinpoint their precise locations within the image. The resulting FROC curve is constructed by plotting the sensitivity of the algorithm, defined as the proportion of true tumors that are both correctly localized and assigned a sufficiently high suspicion score on the y-axis, against the average number of false-positive (FP) detections per case on the x-axis. Unlike the conventional ROC analysis, which considers only the presence or absence of disease, the FROC framework accounts for the possibility of multiple findings per image and provides a more nuanced evaluation of performance in complex diagnostic tasks, such as those encountered in medical imaging where multiple lesions may be present and accurate localization is clinically significant

[31]. In the ROI approach, regions that can contain tumor are defined a priori (e.g., left or right lung) [7].

Regardless of the study design and used methodology, there is one very important rule: to present every metric with an appropriate confidence interval. For the primary performance measure, for example, Se, Sp, and AUC, the CIs are used for decision-making regarding the study success. For secondary and exploratory endpoints, the CIs provide additional information regarding the algorithm performance while not being used in the determination of study success (evaluation of study success criteria). The significance level for the CI calculations must be defined a priori, for example, 5% two-sided. It is crucial to use appropriate CI calculation methods, especially in the presence of multiple observations per single patient, that is, multiple frames per video recording and multiple recordings per patient. Most standard formulas in such a setting are not appropriate as they assume independence of observations, which is not true in the presence of correlated observations. In such a case, Bootstrap confidence intervals are recommended with an appropriate sampling strategy, for example, sampling with replacement only on the patient level to account for correlation between multiple CTs per patient. Failing to take this correlation info into account can result in overoptimistic performance (i.e., too narrow CIs) of the algorithm, which will not be observed in real clinical application in the target population [32,33].

Correlation of observations must also be taken into account during the study design stage, especially in sample size calculation, as failure to consider clustering can result in insufficient sample size and underpowered study, which is a risk for the entire project and can result in excessive cost of algorithm market authorization (additional studies, additional statistical analysis, and delayed release on the market add additional costs). However, not only data clustering can increase the sample size of the study designed to prove the diagnostic performance of the algorithm. Another crucial factor is the number of primary endpoints and the approach in which they are evaluated. In the case of more than one primary endpoint, failing to account for that face in sample size calculations and data analysis can cause so-called multiplicity issues and can drastically increase type I error. There are multiple scenarios for studies with more than one primary endpoint, each requiring different approaches in planning and statistical analysis.

In the case of a diagnostic algorithm, the most common approach is to have two primary endpoints of sensitivity and specificity. Of course, observing success on only one of these would not be enough to declare sufficient performance of the algorithm. Sufficient performance can only be declared if both sensitivity and specificity at the same time meet the prespecified target performance goal, that is, success of the study can be declared only if success on sensitivity and specificity is observed at the same time. In such a case, both endpoints are called co-primary, and a multiplicity issue does not exist; however, this approach inflated type II error. In the most common design, if co-primary endpoints of sensitivity and specificity are used, the power for each should be at least 90% for overall study power to be above 80% (90% x 90% = 80% assuming no positive correlation between endpoints). In general, no more than two co-primary endpoints should be used in the same study, unless the algorithm is designed for more than one distinct type of pathology; however, other approaches can be considered in such a case.

Another approach in dealing with multiple primary endpoints is when demonstration of a diagnostic performance on at least one of the several primary endpoints is sufficient to declare study success. This situation, however very uncommon in the case of diagnostic algorithms, needs a short clarification as it causes a multiplicity issue. Several methods are used to deal with this problem, and all of them involve controlling for type I error inflation by using adjustments like Bonferroni, Holm, or Hochberg methods or by applying hierarchical endpoint testing like the gatekeeping approach.

If the algorithm is designed to identify numerous types of abnormalities, rather than testing each of them separately and thus creating a multiplicity issue, all abnormalities can be combined into one composite endpoint, especially if the prevalence of each abnormality in the target population is small. Using a composite endpoint does create a multiplicity issue; however, it must be clearly justified. Usually in such a scenario, the presence of any of the analyzed types of abnormalities is considered the presence of a composite endpoint, and standard reporting regarding sensitivity and specificity applies. It is a good approach to provide separate results for each of the types of abnormalities; however, do not use them to declare study success and evaluate success criteria (those should use the performance of detection of composite endpoints).

Composite endpoints should not be confused with multicomponent endpoints, which involve several different components occurring on the same patient, for example, ECG can be used to detect signal abnormality like AF but also to estimate left ventricular ejection fraction (LVEF), so multicomponent endpoints can combine patients with AF and low LVEF at the same time. Both composite endpoints and multicomponent endpoints are rare in the case of diagnostic algorithms [34].

Validation study design also involves the decision of whether to include readers that represent human-level performance. Main approaches are:

1. Compare the algorithm's stand-alone performance to predefined target performance goals—no human-level performance involved.
2. Compare algorithm stand-alone performance to human-level performance.
3. Compare algorithm stand-alone performance to the combined algorithm and reader's performance.

The first approach is the simplest and involves the calculation of all required performance metrics by the protocol with appropriate CIs and comparing those CIs to values of TPGs defined in the protocol. The second approach is a special case of the third approach, in which several correlations are by design equal to 0. In this approach, the difference in all required protocol performance metrics is calculated together with appropriate CIs (including CIs for the differences). Then the CIs for the difference are compared to predefined superiority or noninferiority margins to evaluate study success. The third approach is the most complex and is called the multi reader multi case (MRMC) design.

The MRMC design is a robust approach frequently employed in validation studies of diagnostic algorithms. In this framework, multiple readers independently evaluate a set of cases, allowing for comprehensive assessment of both algorithm and human performance across diverse scenarios. The MRMC design helps to quantify variability attributable to both readers and cases, leading to more reliable statistical inference and improved generalizability of study findings.

There are numerous parameters that need to be specified in order to properly plan the MRMC study, but the selection depends on the assumed methodology or analyzing the study results. The two most common approaches are Dorfman–Berbaum–Metz [35] and Obuchowski–Rockette (OR) [36].

The OR method assumes a mixed-effects model for the diagnostic accuracy (e.g., AUC) of a reader j using a particular diagnostic test i on reading occasion q.

$$\hat{\theta}_{ijq} = \mu + \mu_i + r_j + (\mu r)_{ij} + \epsilon_{ijq}$$

Where, $\nu_i = \mu + \mu_i$ is the mean accuracy of diagnostic test i (fixed effect) (and $\Sigma \mu_i = 0$), r_j is a random reader effect reflecting differences in diagnostic accuracy due to different reader abilities (j = 1,...,J), $(\mu r)_{ij}$ is an interaction term for readers and tests (random effect), and ϵ_{ijq} is an error term. $\hat{\theta}_{ijq}$ can be any measure of diagnostic accuracy (e.g., AUC).

In such a model, to calculate power for different sample sizes and numbers of readers, a noncentrality parameter of a noncentral F distribution must be calculated as follows:

$$\hat{\lambda} = \frac{J(\nu_1 - \nu_2)^2}{2(\hat{\sigma}_b^2(1 - \hat{\rho}_b) + (\hat{\sigma}_w^2/Q) + \hat{\sigma}_c^2[(1 - \hat{\rho}_1) + (J - 1)(\hat{\rho}_2 - \hat{\rho}_3)])}$$

Then power can be calculated from:

$$Power = Prob(F_{1,(J-1);\lambda} > F_{1-\alpha;1,(J-1)})$$

where $F_{1,(J-1);\lambda}$ is a random variable having a noncentral F distribution with degrees of freedom of 1 and (J-1) and noncentrality parameter of λ. $F_{1-\alpha;1,(J-1)}$ is the $(1 - \alpha)$ 100th percentile of a central F distribution with degrees of freedom of 1 and (J-1).

Interpretation of the remaining parameters is provided in the table below.

Parameter	Interpretation
α	One-sided significance level
υ_1	Area under the curve (AUC) for the algorithm and readers working together
υ_2	AUC for the readers alone
$\hat{\sigma}_b^2$	The estimated variability in AUC from different readers interpreting the results of the same patients using the same test
$\hat{\sigma}_w^2$	The estimated variability in AUC from the same reader interpreting the results of the same patients using the same test on different occasions

$\hat{\sigma}_c^2$	The estimated variability in AUC from different samples of patients
$\hat{\rho}_1$	The estimated correlation between AUC when the same patients are evaluated by the same reader using different tests
$\hat{\rho}_2 - \hat{\rho}_3$	Difference between the estimated correlation between AUC when the same patients are evaluated by different readers using the same test and the estimated correlation between AUC when the same patients are evaluated by different readers using different tests
$\hat{\rho}_b$	The estimated correlation between AUC when the same readers evaluate patients using different tests
J	Number of readers
Q	Number of times each reader interprets the test results of each patient using the same test

There are several types of MRMC designs commonly employed in the validation of diagnostic algorithms. The most basic form is the fully crossed design, in which every reader evaluates every case, allowing for comprehensive assessment of both reader and case variability. Another variant is the split-plot design, where readers are assigned to evaluate only subsets of cases, which can increase efficiency in large studies but may require more complex statistical analysis. Additionally, the paired-reader design involves each case being read by a pair of readers, facilitating direct comparison and minimizing workload. These design types enable robust statistical inference and help to address the variability inherent in reader interpretations and case presentations, ultimately leading to more reliable validation results for diagnostic algorithms.

Each type of MRMC design will determine the relationship between values of correlations defined in the table above. If both patients and readers are paired, all correlations must be positive or 0, and the following relation must be true: $\hat{\rho}_1 \geq \hat{\rho}_2 \geq \hat{\rho}_3$. If the patients are unpaired both $\hat{\rho}_1$ *and* $\hat{\rho}_3$ are equal exactly 0. If patients are paired but readers are not, $\hat{\rho}_1$ must be equal to $\hat{\rho}_3$ but can be positive. In case both patients and readers are unpaired, both $\hat{\rho}_1$ and $\hat{\rho}_3$ must be 0. In a special case when readers are paired, and patients are paired but only within a reader, both $\hat{\rho}_2$ and $\hat{\rho}_3$ must be 0.

This illustrates that, although MRMC studies can be methodologically complex and demanding to both plan and analyze, their primary strength lies in their exceptional flexibility and realism. The MRMC framework is capable of accommodating a wide range of clinically relevant scenarios by explicitly modeling multiple sources of variability, including differences between readers, heterogeneity among patients or cases, and the interactions between readers and diagnostic tests. By accounting for these variance components, MRMC designs provide statistically valid and generalizable estimates of diagnostic performance that more closely reflect real-world clinical practice than simpler validation approaches.

Importantly, this flexibility allows investigators to tailor the study design to practical constraints and scientific objectives, such as the available number of expert readers, the size and composition of the case set, and the intended clinical use of the algorithm. Researchers can choose between fully crossed, partially crossed, or other designs to balance statistical power, feasibility, and cost. As a result, MRMC studies enable informed trade-offs between methodological rigor and operational efficiency, empowering investigators to select the most powerful and feasible design while preserving robust inference about algorithm performance relative to human readers or combined human–AI workflows.

8.7.1 The convergence of artificial intelligence and spatial computing

As medical imaging moves from 2D slices to high-fidelity 3D and 4D representations, the interface between the clinician and the data must also evolve. The traditional paradigm of viewing volumetric data on flat 2D monitors creates a cognitive bottleneck, forcing the physician to mentally reconstruct the spatial anatomy. The future of diagnostic support lies in the integration of AI-driven segmentation with mixed reality (MR) and spatial computing. Emerging platforms allow for the projection of segmented anatomical models directly onto the patient's body in the operating room (augmented reality) or facilitate immersive surgical planning in a virtual space. In this context, AI serves as the backend engine that automatically segments and classifies tissues, while MR serves as the frontend interface, allowing clinicians to "manipulate" the data intuitively. This synergy promises to reduce cognitive load and improve the precision of complex interventions, such as neurosurgery or orthopedic reconstruction.

8.7.2 Fragmentation of data sources

Currently, data acquisition remains largely a localized effort, dependent on bilateral agreements between individual research groups and hospitals. This fragmentation creates "data silos" and limits the scalability of AI solutions. A major future challenge is the establishment of country-level and international infrastructures that streamline data acquisition. The goal is to move from ad-hoc data transfers to federated networks where standardized, pre-anonymized data streams are accessible to certified researchers. Initiatives aiming to unify these disparate sources are critical for training "Foundation Models" in medicine, which require data diversity far exceeding the capacity of any single institution.

8.7.3 EDITH and the vision of the virtual human twin

The ultimate manifestation of this technological convergence is the VHT—an integrated, multiscale, and dynamic digital representation of the patient. The VHT goes beyond static anatomy; it incorporates physiological rules, genomic data, and real-time sensor inputs to simulate the patient's health trajectory. Recognizing the immense potential and complexity of this goal, the European Commission has launched EDITH (an ecosystem for digital twins in healthcare) [37]. This flagship coordination action acts as a roadmap to foster an inclusive ecosystem for VHTs in Europe. EDITH addresses the critical fragmentation of current research by establishing standards for data exchange, model validation, and ethical compliance. The initiative aims to create a federated repository where organ-specific twins (e.g., a "Digital Heart" or "Digital Brain") can be interconnected to model systemic interactions. For the medical community, the success of EDITH represents a transition to in silico medicine, where treatments can be optimized on a digital avatar before being administered to the patient, minimizing risk and maximizing efficacy.

8.7.4 The European Health Data Space as the backbone

None of these advancements are possible without data. The fuel for training robust AI models and constructing accurate VHTs is high-quality, diverse, and accessible data. The upcoming EHDS [38] serves as the infrastructural backbone for these future directions. By creating a unified framework for the secondary use of health data across EU member states, the EHDS will unlock vast datasets that are currently siloed in national hospitals. This will enable the training of models on millions of cases rather than hundreds, drastically improving the generalization capabilities of AI systems and ensuring that digital twins are representative of the diverse European population.

We stand at an inflection point in digital medicine. The foundational technologies—deep learning architectures, 3D acquisition protocols, and validation frameworks—are maturing. The challenge for the next decade is integration. By aligning technical innovation with strategic initiatives like EDITH and the EHDS, the medical data analysis community is poised to deliver on the promise of truly personalized, predictive, and precision medicine.

8.8 Future directions

The trajectory of medical data analysis is moving decisively away from isolated, task-specific tools toward holistic, integrated ecosystems. While the current state-of-the-art focuses on optimizing individual models for segmentation or classification, the future lies in the convergence of these algorithms with high-performance computing, spatial visualization, and interoperable data infrastructures. This synthesis aims to transcend the limitations of current "decision support" to create fully immersive, predictive environments for personalized medicine.

References

[1] Joshi G, Jain A, Araveeti SR, Adhikari S, Garg H, Bhandari M. FDA-approved artificial intelligence and machine learning (AI/ML)-enabled medical devices: an updated landscape. Electronics 2024;13:498. Available from https://doi.org/10.3390/electronics13030498.

[2] Clinical decision support software guidance for industry and food and drug administration staff document issued on January 6, 2026. https://www.fda.gov/regulatory-information/search-fda-guidance-documents/clinical-decision-support-software.

[3] https://www.fda.gov/medical-devices/software-medical-device-samd/artificial-intelligence-enabled-medical-devices

[4] https://www.fda.gov/science-research/artificial-intelligence-and-medical-products/fda-digital-health-and-artificial-intelligence-glossary-educational-resource.

[5] Vokinger KN, Feuerriegel S, Kesselheim AS. Continual learning in medical devices: FDA's action plan and beyond. The Lancet Digital Health 2021;3(6):e337–8.

[6] Dendukuri N, Joseph L. Bayesian approaches to modeling the conditional dependence between multiple diagnostic tests. Biometrics 2001;57(1):158–67. Available from https://doi.org/10.1111/j.0006-341x.2001.00158.x. PMID: 11252592.

[7] Zhou XH, Sun J, Pennello GA, Obuchowski NA, McClish DK. Statistical methods in diagnostic medicine. John Wiley & Sons; 2026. Available from 10.1002/9780470906514.

[8] https://www.rsna.org/.

[9] Wagner P, Strodthoff N, Bousseljot R, Samek W, Schaeffter TPTB-XL. A large publicly available electrocardiography dataset (version 1.0.3). PhysioNet 2022. Available from https://doi.org/10.13026/kfzx-aw45.

[10] X. Wang, Y. Peng, L. Lu, Z. Lu, M. Bagheri, and R.M. Summers. Chestx-ray8: hospital-scale chest x-ray database and benchmarks on weakly-supervised classification and localization of common thorax diseases. In: Proceedings of the IEEE conference on computer vision and pattern recognition (CVPR). 2017. p. 3462–71.

[11] Irvin J, Rajpurkar P, Ko M, Yu Y, Ciurea-Ilcus S, Chute C, et al. Chexpert: a large chest radiograph dataset with uncertainty labels and expert comparison. In: Proceedings of the AAAI conference on artificial intelligence. 2019.

[12] Johnson AE, Pollard TJ, Berkowitz SJ, Greenbaum NR, Lungren MP, Deng CY, et al. MIMIC-CXR: a de-identified publicly available database of chest radiographs with free-text reports. Scientific Data 2019;6(1):317.

[13] Bustos A, Pertusa A, Salinas JM, De La Iglesia-Vaya M, et al. Padchest: a large chest x-ray image dataset with multi-label annotated reports. Medical Image Analysis 2020;66:101797.

[14] Zeng B, Yin Y, Liu Z. Understanding bias in large-scale visual datasets. In: The thirty-eighth annual conference on neural information processing systems 2024.

[15] Ashby D. Bayesian statistics in medicine: a 25 year review. Statistics in Medicine 2006;25:3589–631. Available from https://doi.org/10.1002/sim.2672.

[16] He, K, Zhang, X, Ren, S, Sun, J. Deep residual learning for image recognition. In: Proceedings of the IEEE conference on computer vision and pattern recognition. 2016. p. 770–8.

[17] Ronneberger O, Fischer P, Brox T. U-net: convolutional networks for biomedical image segmentation. International conference on medical image computing and computer-assisted intervention. Cham: Springer international publishing; 2015. p. 234–41.

[18] Isensee F, Petersen J, Klein A, Zimmerer D, Jaeger PF, Kohl S, et al. nnu-net: self-adapting framework for u-net-based medical image segmentation. arXiv 2018. 1809.10486.

[19] Redmon J., Divvala S., Girshick R., Farhadi A. You only look once: unified, real-time object detection. In: Proceedings of the IEEE conference on computer vision and pattern recognition. 2016. p. 779–88).

[20] Dosovitskiy A. An image is worth 16x16 words: Transformers for image recognition at scale. arXiv preprint arXiv:2010. 11929. 2020. Jieneng Chen, Jieru Mei, Xianhang Li, Yongyi Lu, Qihang Yu, Qingyue Wei, Xiangde Luo, Yutong Xie, Ehsan Adeli, Yan Wang, Matthew P. Lungren, Shaoting Zhang, Lei Xing, Le Lu, Alan Yuille, Yuyin Zhou.

[21] TransUNet: rethinking the U-Net architecture design for medical image segmentation through the lens of transformers. Medical Image Analysis, Volume 2024;97:103280. https://doi.org/10.1016/j.media.2024.103280.

[22] Hatamizadeh A, Tang Y, Nath V, Yang D, Myronenko A, Landman B, et al. Unetr: transformers for 3d medical image segmentation. In: Proceedings of the IEEE/CVF winter conference on applications of computer vision 2022:574–84.

[23] Baumgartner CF, Koch LM, Pollefeys M, Konukoglu E. An exploration of 2D and 3D deep learning techniques for cardiac MR image segmentation. International workshop on statistical atlases and computational models of the heart. Cham: Springer International Publishing; 2017. p. 111–9.

[24] Goodfellow IJ, Pouget-Abadie J, Mirza M, Xu B, Warde-Farley D, Ozair S, et al. Generative adversarial nets. Advances in Neural Information Processing Systems 2014;27.

[25] Ho J, Jain A, Abbeel P. Denoising diffusion probabilistic models. Advances in Neural Information Processing Systems 2020;33:6840–51.

[26] Sudre CH, Li W, Vercauteren T, Ourselin S, Jorge Cardoso M. Generalised dice overlap as a deep learning loss function for highly unbalanced segmentations. International Workshop on Deep Learning in Medical Image Analysis. Cham: Springer International Publishing; 2017. p. 240–8.

[27] Lin T.Y., Goyal P., Girshick R., He K., Dollár P. Focal loss for dense object detection. In: Proceedings of the IEEE international conference on computer vision. 2017. p. 2980–8.

[28] Selvaraju R.R., Cogswell M., Das A., Vedantam R., Parikh D., Batra D. Grad-cam: visual explanations from deep networks via gradient-based localization. In: Proceedings of the IEEE international conference on computer vision. 2017. p. 618–26.

[29] Lundberg SM, Lee SI. A unified approach to interpreting model predictions. Advances in Neural Information Processing Systems 2017:30.

[30] He X, Frey E. ROC, LROC, FROC, AFROC: an alphabet soup. Journal of the American College of Radiology JACR 2009;6(9):652–5. Available from https://doi.org/10.1016/j.jacr.2009.06.001. PMID: 19720362.

[31] Bandos AI, Rockette HE, Song T, Gur D. Area under the free-response ROC curve (FROC) and a related summary index. Biometrics 2009;65:247–56. Available from https://doi.org/10.1111/j.1541-0420.2008.01049.x.

[32] Ren S, Lai H, Tong W, Aminzadeh M, Hou X, Lai S. Nonparametric bootstrapping for hierarchical data. Journal of Applied Statistics 2010;37(9):1487–98. Available from https://doi.org/10.1080/02664760903046102.

[33] Field CA, Welsh AH. Bootstrapping clustered data. Journal of the Royal Statistical Society Series B: Statistical Methodology 2007;69:369–90. Available from https://doi.org/10.1111/j.1467-9868.2007.00593.x.

[34] GUIDANCE DOCUMENT. Multiple endpoints in clinical trials, guidance for industry. 2022.

[35] Dorfman DD, Berbaum KS, Metz CE. Receiver operating characteristic rating analysis. Generalization to the population of readers and patients with the jackknife method. Investigative Radiology 1992;27:723–31.
[36] Obuchowski NA, Rockette HE. Hypothesis testing of diagnostic accuracy for multiple readers and multiple tests: an ANOVA approach with dependent observations. Communications in Statistics–Simulation and Computation 1995;24:285–308.
[37] https://www.edith-csa.eu/.
[38] https://health.ec.europa.eu/ehealth-digital-health-and-care/european-health-data-space-regulation-ehds_pl.

CHAPTER

Conclusions

9

Artur Rydosz[1,2], Klaudia Proniewska-van Dam[2,3] and Peter van Dam[2,4]

[1]*Biomarkers Analysis LAB, Institute of Electronics, AGH University of Kraków, Kraków, Poland,*
[2]*Functional and Virtual Medical 3D Imaging Laboratory, Department of Diagnostic Imaging, University Hospital in Kraków, Kraków, Poland,*
[3]*Center for Digital Medicine and Robotics, Jagiellonian University Medical College, Kraków, Poland,*
[4]*Department of Radiology, Jagiellonian University Medical College, Kraków, Poland*

"Nothing in life is to be feared, it is only to be understood. Now is the time to understand more, so that we may fear less."

—Maria Skłodowska-Curie

The dynamic development of functional and virtual 3D imaging in cardiovascular medicine is redefining contemporary standards in diagnostics, therapeutic planning, and clinical education. By integrating advanced imaging modalities with hemodynamic analysis, computational simulations, and immersive virtual environments and 3D prints, clinicians can now not only visualize anatomical structures but also understand their functional behavior in unprecedented ways. Imaging data are transformed into powerful decision-support tools that enhance procedural precision, enable personalized treatment strategies, and improve patient safety. As a result, daily clinical practice becomes increasingly data-driven, predictive, and tailored to each patient's cardiovascular profile.

Equally significant is the educational impact of these technologies. Interactive 3D models, virtual simulations, and procedure-specific rehearsal platforms create a transformative learning ecosystem for medical students, trainees, and multidisciplinary teams. Immersive visualization shortens the learning curve, allows for safe repetition of complex interventions, and deepens comprehension of anatomical and physiological relationships. Education in functional and virtual 3D imaging, therefore, becomes not merely a supplementary skill but a foundational component in preparing healthcare professionals to implement innovative cardiovascular therapies effectively and responsibly.

Looking ahead, the future of cardiovascular medicine will be increasingly shaped by the synergy between 3D imaging, artificial intelligence, and precision medicine with a personalized approach. Successful integration of these advancements requires not only technological infrastructure but also structured training,

Functional and Virtual Medical 3D Imaging in Cardiovascular Medicine. DOI: https://doi.org/10.1016/B978-0-443-44570-5.00001-X

standardization, and sustained collaboration between clinicians, engineers, and researchers. Functional and virtual medical 3D imaging is no longer a concept of the future—it is an evolving clinical reality reshaping how we diagnose, treat, and educate in everyday cardiovascular practice.

Index

Note: Page numbers followed by *f* indicate figures, *t* indicate tables.

Y

Printed and bound by CPI Group (UK) Ltd, Croydon, CR0 4YY
14/09/2026
02216297-0003